INVERSE AND ILL-POSED PROBLEMS SERIES

Inverse Problems of Wave Processes

Also available in the Inverse and Ill-Posed Problems Series:

Inverse Problems for Kinetic and other Evolution Equations
Yu.E. Anikonov
Uniqueness Problems for Degenerating Equations and Nonclassical Problems
S.P. Shishatskii, A. Asanov and E.R. Atamanov
Uniqueness Questions in Reconstruction of Multidimensional Tomography-Type Projection Data
V.P. Golubyatnikov
Monte Carlo Method for Solving Inverse Problems of Radiation Transfer
V.S. Antyufeev
Introduction to the Theory of Inverse Problems
A.L. Bukhgeim
Identification Problems of Wave Phenomena - Theory and Numerics
S.I. Kabanikhin and A. Lorenzi
Inverse Problems of Electromagnetic Geophysical Fields
P.S. Martyshko
Composite Type Equations and Inverse Problems
A.I. Kozhanov
Inverse Problems of Vibrational Spectroscopy
A.G. Yagola, I.V. Kochikov, G.M. Kuramshina and Yu.A. Pentin
Elements of the Theory of Inverse Problems
A.M. Denisov
Volterra Equations and Inverse Problems
A.L. Bughgeim
Small Parameter Method in Multidimensional Inverse Problems
A.S. Barashkov
Regularization, Uniqueness and Existence of Volterra Equations of the First Kind
A. Asanov
Methods for Solution of Nonlinear Operator Equations
V.P. Tanana
Inverse and Ill-Posed Sources Problems
Yu.E. Anikonov, B.A. Bubnov and G.N. Erokhin
Methods for Solving Operator Equations
V.P. Tanana
Nonclassical and Inverse Problems for Pseudoparabolic Equations
A. Asanov and E.R. Atamanov
Formulas in Inverse and Ill-Posed Problems
Yu.E. Anikonov
Inverse Logarithmic Potential Problem
V.G. Cherednichenko
Multidimensional Inverse and Ill-Posed Problems for Differential Equations
Yu.E. Anikonov
Ill-Posed Problems with A Priori Information
V.V. Vasin and A.L. Ageev
Integral Geometry of Tensor Fields
V.A. Sharafutdinov
Inverse Problems for Maxwell's Equations
V.G. Romanov and S.I. Kabanikhin

INVERSE AND ILL-POSED PROBLEMS SERIES

Inverse Problems of Wave Processes

A.S. Blagoveshchenskii

///VSP///
UTRECHT • BOSTON • KÖLN • TOKYO
2001

VSP
P.O. Box 346
3700 AH Zeist
The Netherlands

Tel: +31 30 692 5790
Fax: +31 30 693 2081
vsppub@compuserve.com
www.vsppub.com

First published in 2001

ISBN 90-6764-344-0

Printed in The Netherlands by Ridderprint bv, Ridderkerk.

Contents

Introduction

This monograph is devoted to inverse problems in the theory of wave processes.

Problems of mathematical physics are conventionally classified as direct or inverse problems according to the following rule. Let us consider a process in a physical system subjected to external sources. The properties of the system are supposed to be known. Then the problem of describing the process is referred to as the direct problem. However, another case is possible. Suppose that we have an additional information about how the process operates, but we do not know some parameters of the system or the sources. The problem of recovering these parameters is called the inverse problem.

We shall focus our attention on a special class of inverse problems when the physical process in question is the propagation of nonstationary waves. The following situation is typical. Suppose that an inhomogeneous wave-conducting medium fills a domain Ω. Wave sources are situated outside Ω or on its boundary. The waves generated by these sources are propagated inside Ω and, interacting with medium nonhomogeneities, are scattered. The scattered waves are registered by receivers located outside Ω. Having got these data, it is required to find a function (or a set of functions) describing the properties of the inhomogeneous medium. It is reasonable to consider the following questions. Is the information sufficient to recover the desired parameters (the uniqueness theorem)? If so, which is the algorithm for their reconstruction? If the information is not sufficient, which additional information should be available? Does the additional information lead to narrowness of the class of admissible models of the medium, or it may be given in terms of the wave field observed? Which is the class of all possible data fitting the chosen model of the medium? All the questions in various specific cases are considered through the book.

It should be noted that in this monograph we study only dynamical inverse problems, i. e., such problems whose data are the values of wave fields. We do not include the so-called kinematic inverse problems whose data are only the times waves take to run through the medium.

The problems considered in this book find extensive application in such fields as geophysics, acoustics, elasticity theory, electrodynamics, etc. They are widely covered in the literature. We have no intention to provide a comprehensive review of the literature devoted to the subject. Indeed, with the rapid growth of the theory of inverse problems and its relationship to many branches of natural sciences, such a task is rather difficult. The aim of this book is to present our own results in this field. We refer those who wish to examine the subject more closely to Lavrent'ev *et al.* (1986), Romanov (1987), Yakhno (1985, 1990), Goryunov and Saskovets (1989), Nizhnik (1991), Romanov and Kabanikhin (1991). Recent results concerning inverse problems for hyperbolic equations and numerical methods of their solution can be found in Romanov and Kabanikhin (1994), Kabanikhin and Lorenzi (1999).

In mathematical terms, the monograph deals with the problem of determination of one or more coefficients of a hyperbolic equation or a system of hyperbolic equations. The desired coefficients are functions of point. Most attention is concentrated on the case when the required functions depend only on one coordinate. A distinction needs to be drawn between the case when waves propagate in a one-dimensional medium (it means that the functions describing the wave field depend only on one spatial variable and time) and the case when the wave field depends on many spatial variables. In the former case, we call the inverse problems one-dimensional; while in the latter case, we refer to them as being in layered media. In the literature, inverse problems in both the cases are often called one-dimensional. It should be noted, however, that though the inverse problems for layered media are usually reduced to one-dimensional ones in our terminology, they have peculiar features. This is the reason why we separate them from one-dimensional problems. Inverse problems may also be classified according to the types of wave fields. If the wave field is scalar, the wave process is described by one partial differential equation. In the case of vector (or tensor) wave field, we deal with a system of such equations.

Taking into account the preceding, we have organized this book in three chapters. Chapter 1 is devoted mostly to methods of solution of one-dimensional inverse problems. These methods are at the heart of the further discussion. Note that the distinctive feature of the presented methods is their local character. In order to clear up the matter, let us consider the following typical problem.

Suppose that the wave propagation is described by the equation

$$u_{tt} = u_{xx} + q(x)u, \qquad x > 0, \tag{1}$$

where $u(x,t)$ satisfies the initial and boundary conditions

$$u|_{t<0} \equiv 0, \quad u|_{x=0} = F(t), \quad u_x|_{x=0} = G(t). \tag{2}$$

As is known, given the coefficient $q(x)$, we can find the wave field $u(x,t)$ from one of the functions $F(t)$, $G(t)$. Therefore, if $q(x)$ is known, setting both the functions $F(t)$ and $G(t)$ causes the problem to become overdetermined: if $F(t)$ and $G(t)$ are arbitrary, then the problem may be unsolvable for some $q(x)$. So, taking $F(t)$ and $G(t)$, we thereby impose some restrictions on the choice of $q(x)$. Hence, it is reasonable to think of $q(x)$ as unknown and to recover it from $F(t)$ and $G(t)$.

The following statement turns out to be true: *the function $q(x)$ is uniquely determined from the functions $F(t)$ and $G(t)$.*

One of characteristic properties of equation (1) is finiteness of the speed of wave propagation (more precisely, this speed is equal to unity). This means that the wave emitted at $t = 0$ by a source located at $x = 0$ can reach the point x_0 in the time $t = x_0$. Due to the medium nonhomogeneity or, in other words, since $q(x)$ is not constant, the wave is scattered. The wave scattered in a neighborhood of the point x_0 gets an observer being at the point $x = 0$ by the time $t = 2x_0$. Therefore, fixing the scattered waves in the time interval $(0, 2x_0)$, we can obtain the information about how the coefficient $q(x)$ changes in the interval $(0, x_0)$. If we observe the scattered wave for $t \in (0, 2x_1)$, $x_1 > x_0$, we get an additional information on $q(x)$ for $x > x_0$. This information is unnecessary if our intention is to recover the coefficient $q(x)$ only in the interval $x \in (0, x_0)$. Therefore, it would appear natural that to recover the coefficient $q(x)$ in the interval $(0, x_0)$ it suffices to know the data of the inverse problem only for $t \in (0, 2x_0)$. Really, this holds in the problem (1), (2) and in many similar inverse problems.

We shall develop two methods for solving the inverse problem. Either of the methods is local in the sense that it takes into account the above property.

Both the methods use finiteness of the wave speed to reduce the inverse problem to one or more integral equations. To be more precise, using the first method, we reduce the inverse problem to a nonlinear system of second-order integral equations of Volterra type (or of a similar type). The second method allows us to reduce the inverse problem to a second-order linear

Fredholm integral equation of Gel'fand – Levitan – Krein – Marchenko type (see Gel'fand and Levitan, 1951; Krein, 1955; Marchenko, 1955). For brevity, the first method will be referred to as nonlinear; and the second method, as linear. Comparing these methods, we can say that when both the methods can be applied (for example, in the problem (1), (2)), the linear method yields better results. Using the linear method, we can investigate the inverse problem completely, that is, we can prove existence and uniqueness theorems and, in certain cases, construct even an explicit solution. By the nonlinear method, we can demonstrate, as a rule, only a uniqueness theorem. The existence of the solution can be established only in the small. Moreover, since the system of integral equations to which the inverse problem is reduced is not always of Volterra type, it may be that the uniqueness theorem holds only in the small and the existence theorem fails at all. (Here, a theorem is said to be true in the small if it holds on an interval $(0, l)$ dependent on the data of the considered problem). However, the nonlinear method can be applied to a wider class of inverse problems.

These methods are discussed in Sections 1.3, 1.4, 1.6 which hold a central position in Chapter 1. The first two sections of the book can be regarded as an introduction. In these sections we pose direct and inverse problems, discuss their physical sense, and study some properties of solutions of the direct problems. The method presented in Section 1.4 is adapted to the case when desired coefficients of equations are continuous. In Section 1.5 this method is extended to the case of discontinuous coefficients.

As mentioned above, in some special cases, inverse problems can be solved explicitly in terms of elementary functions. This subject is covered in Sections 1.7, 1.8. In Section 1.7 we give an example of inverse problems which admit an explicit solution. In Section 1.8 we relate the inverse problems to a class of nonlinear systems of ordinary differential equations. This relation allows us, on the one hand, to construct explicit solutions of the systems of equations and, on the other hand, to use known approximate methods of solution of differential equations for numerical solution of the inverse problems.

In Section 1.9, we study the inverse problem in a more general case when the coefficients $q_1(x)$, $q_2(x)$ of the hyperbolic system

$$\psi_{1t} + \psi_{1x} = q_1\psi_2,$$
$$\psi_{2t} - \psi_{2x} = q_2\psi_1$$

are to be determined. It is shown, in particular, how the linear method should be modified to be applicable to the case. The uniqueness and existence theorems are also proved.

Chapter 2 is devoted to scalar inverse problems of wave propagation in a layered medium. The stratification of the medium, i. e., the dependence of the coefficients describing properties of the medium only on one coordinate, makes it possible to apply the method of separation of variables in one or another form. As a result, we obtain a one-dimensional inverse problem which is solvable by the methods of Chapter 1. Solving this problem, we find a function depending on the coefficients of the initial multidimensional equation and on parameters (separation constants). Varying these parameters, we recover, if possible, all the coefficients of the initial equation or show that they can not be determined uniquely.

In Section 2.1 we study inverse acoustic problems in a layered medium, i. e., problems for the equation

$$au_{tt} = \operatorname{div}(b \operatorname{grad} u) \tag{3}$$

where a and b are functions of one Cartesian coordinate z (the vertical variable). In particular, the problems are investigated in the case when waves are propagated in the half-space $z > 0$, and the point source generating the waves is situated on the boundary. Data of the inverse problem can be set in different ways. We consider the following two variants.

A. The values of the function $\tilde{u}(k, 0, t)$ are given for two values of the parameter k. Here $\tilde{u}(k, z, t)$ is the Fourier transform of $u(x, z, t)$ with respect to the horizontal variables x, and k is the variable dual to x.

B. We consider the Radon transformation of the function u as a function of x and t, that is, the function which to a plane in the space of x and t associates the integral of u over the plane. The values of the Radon transformation of $u(x, 0, t)$ are given for two fixed values of the vector normal to the plane (we choose spatially oriented planes).

Another problem, the so-called inverse scattering problem, arises when waves, generated by a plane wave incident from the homogeneous half-space $z > 0$, are propagated in the whole space of variables (x, z). We study the problem in the case when the scattered waves are fixed for two different directions of the incident wave.

In this section we also consider the following situations. 1) The waves are propagated in a wave guide (the variable x belongs to a bounded domain Ω, and self-adjoint boundary conditions hold on the boundary of Ω). 2) The waves are propagated in a ball (the coefficients a and b depend on the radial variable). In either of these cases the variables are separable. We assume that the functions $f_k(t) = (u, \varphi_k)|_{z=0}$ are known for two values of k, where φ_k

are eigenfunctions of the operator Δ_x in Ω or of the Beltrami–Laplace operator on the sphere.

It is shown that in each of the above formulations the data of the inverse problem is sufficient for unique recovery of both the coefficients a and b in equation (3).

Data of the inverse problem for equation (3) in a half-space can also be given by two moments of the function $u(x, 0, t)$:

$$\int u(x,0,t)|x|^m \, dx, \qquad m = 0, 2.$$

It turns out that in this case the problem is reduced to two one-dimensional inverse problems which should be solved successively. The first problem is a special case of the inverse problem considered in Chapter 1. The second is reduced to a linear integral Volterra equation of second kind. It is remarkable that this equation can be solved explicitly in quadratures. An analogous result can be obtained in the scattering problem if the known reflected waves are generated by incident waves whose angles of incidence are close.

In Sections 2.2 and 2.3 we deal with the inverse problems for the equation

$$u_{tt} = Lu.$$

Here L is a second-order elliptic operator in $\mathbb{R}^n$ with coefficients dependent on the vertical coordinate z. In Section 2.2 the problem is considered in the half-space $z > 0$ with a point source on the boundary $z = 0$. In Section 2.3 we study the problem of scattering of a plane wave incident from a homogeneous half-space (if the half-space is $z < 0$, L is the Laplace operator). It is proved that even if we have the most possible information on the inverse problem data: $u|_{z=0}$ are known at all the points of the boundary $z = 0$ for all t (the problem in the half-space) or scattered waves are measured for all possible directions of incident waves (the scattering problem), it is impossible to recover all the coefficients of the operator L. We can uniquely determine only some functions expressed in terms of the coefficients. The number of these functions is much less than that of the coefficients. For example, in the three-dimensional case, the number of the coefficients is 10, while the number of the functions which can be determined is 6. This result is rather unexpected because the problem is overdetermined. Consider, for example, the problem in the half-space in the three-dimensional case. Given a function of three variables (time and two horizontal coordinates), that is, a continuum set of functions of one variable, we have to find 10 functions of one variable. Nevertheless, we fail to determine uniquely the 10 functions.

In Chapter 3 we deal with inverse problems for such system of hyperbolic equations as elasticity equations in stratified media and the acoustic equations for moving media.

In Section 3.1 we study the Lamb problem, important in geophysics. This is the problem of vibrations of a stratified elastic half-space in response to an instantaneous attack directed towards a point on the boundary. It is shown that if the impact is directed obliquely to the surface, then, by observing the surface vibrations, we can recover uniquely the medium parameters depending on the depth: the density and the Lamé parameters λ and μ.

In Sections 3.2 and 3.3 we study inverse problems of acoustics for a moving medium. In Section 3.2 we consider the case when all the medium parameters, including the vector of flow velocity in the medium, are functions of only one Cartesian coordinate z; and the vector of flow velocity is directed perpendicularly to the z-axis. We suppose also that plane waves are incident from the homogeneous immovable half-space upon the medium and reflected waves are known for five different angles of incidence, satisfying an additional condition in the form of inequality. Then it is found that all the acoustic characteristics of the medium are uniquely determined.

In Section 3.3 we study a one-dimensional inverse problem in the case when the acoustic field depends only on the coordinate $z \in (0, \infty)$ and the vector of flow velocity is directed along the z-axis. The wave field is generated by a source located at the point $z = 0$. It is required to recover the acoustic characteristics of the medium for $z > 0$ from the wave field measured for $z = 0$. It turns out that all the parameters of the medium can not be reconstructed. It is possible to find only one or two characteristics if the other are known. It should be noted that if the desired function is the flow velocity, then the problem is reduced to a specific system of integral equations which involves the integration operators of an unknown function along curves depending on the unknown function as well.

In Section 3.4 we consider inverse problems for a one-dimensional first-order hyperbolic system of the general form. The problems are to recover some coefficients of the system from the others.

In Section 3.5 we study an inverse problem for the system of second-order integral equations

$$C^{-2}U_{tt} - U_{xx} = QU,$$

where C, Q, U are 2×2 matrices,

$$C = \begin{pmatrix} 1 & 0 \\ 0 & c \end{pmatrix}, \qquad 0 < c < 1, \qquad Q = Q(x).$$

The following information is assumed to be known:

$$U|_{t<0} = 0, \quad U|_{x=0} = \delta(t)E, \quad V|_{x=0} = (C^{-1}U_t + U_x)|_{x=0} = F(t),$$

where E is the unit matrix. The problem is to recover $Q(x)$ from the given $F(t)$. It is shown that if the matrix $F(t)$ is given in the interval $(0, 2T)$, then $Q(x)$ can not be uniquely recovered in the interval $(0, T)$. For $Q(x)$ to be uniquely determined it suffices to know the values of some components of $Q(x)$ in certain intervals (different for different components) adjoining the interval $(0, T)$ from the right. We also consider the above inverse problem in the case when $Q(x) \equiv 0$ in an interval $(l, +\infty)$. It is shown that in this case the matrix $Q(x)$ can be recovered uniquely from the matrix $F(t)$. The existence theorem is proved for an arbitrary continuous $F(t)$ such that its first column is given in the interval $(0, l(1+1/c))$ and the second in $(0, 2l/c)$. Here $2l/c$ is the time a wave takes to propagate the distance l twice, moving with velocity c; and $l(1 + 1/c)$ is the time it takes for a wave to run from 0 to l with velocity c and back with velocity 1. All results of this paragraph are valid in the small.

When writing this book, the author was supported by Swedish Academy of Sciences (KVA) (project 1414, 1994–1996). The work was partially supported by Russian Foundation for Basic Researches grants No. 98–01–00314 and No. 99–02–16844.

The book was initiated by Prof. M. I. Belishev and Prof. S. I. Kabanikhin. The author is grateful to them for the initiative.

Chapter 1.

One-dimensional inverse problems

1.1. SETTING OF A PROBLEM FOR STRING EQUATION

1.1.1. General remarks

This chapter is devoted to methods of solution of inverse problems for the string equation and related problems. We suggest two methods below. First of them is based on application of the nonlinear Volterra systems of integral equations. This method seemed to be first proposed by Blagoveshchenskii (1966), and later it was widely used by many authors (see, for example, Romanov, 1987). The second method applies linear integral equations. Such equations for inverse problems were first derived by Gel'fand and Levitan (1951), Krein (1955), Marchenko (1955). Note that our considerations are conceptually more close to the papers of Blagoveshchenskii (1971), Kay and Moses (1956), and Pariiskii (1969). If we compare these methods in the case when we can apply both of them we find that the second is better because it gives more complete results. However, the first method has a wider domain of application: it can be successfully used for solving some inverse problems (not only for the string equation), where the second method fails.

1.1.2. Mathematical setting of the problem

The equation of string vibration (with no absorption) is as follows

$$\rho u_{tt} = \frac{\partial}{\partial x}(T u_x) \tag{1.1.1}$$

where $\rho > 0$ is the string density; $T > 0$ is the string tension; ρ and T are assumed to depend on x. Suppose that equation (1.1.1) holds for $x > 0$, and the string is static for $t < 0$:[1]

$$u|_{t<0} = 0. \tag{1.1.2}$$

Suppose we know the excursion of the point $x = 0$ of the string from the equilibrium position

$$u|_{x=0} = F(t) \tag{1.1.3}$$

and the force acting at this point

$$Tu_x|_{x=0} = G(t). \tag{1.1.4}$$

The classes of the given and desired functions will be discussed further. It is required to recover either one of the functions $\rho(x)$, $T(x)$ (the other is thought of as known), or some their combination (for example, the speed of wave propagation $c = \sqrt{T/\rho}$, or the acoustic rigidity $\sigma = \sqrt{T\rho}$), given $F(t)$ and $G(t)$. For definiteness, we shall assume that the acoustic rigidity is the desired function. Indeed, having found this function (considering that either ρ, T, or c is known), we can easily recover the coefficients of equation (1.1.1). For example, if c is known, and σ is found, then $T = \sigma c$, $\rho = \sigma/c$.

1.1.3. Physical interpretation of inverse problem data

Data (1.1.3), (1.1.4) of the inverse problem arise in the following physical situations:

1) The point $x = 0$ is the end of the string. Condition of the fixed end

$$\left(T\frac{\partial u}{\partial x} - hu\right)\Big|_{x=0} = G_1(t) \tag{1.1.5}$$

and the motion of the end

$$u|_{x=0} = F(t) \tag{1.1.6}$$

are given.

Condition (1.1.5) means that the end $x = 0$ is elastically fixed (the constant h is the coefficient of elastic fixing; if $h = 0$, the end is free) and the external force $G_1(t)$ acts at this end.

[1] Instead of the initial condition $u|_{t<0} = 0$, the following equivalent condition $u|_{t=0} = u_t|_{t=0} = 0$ is commonly accepted.

We assume that the coefficient h is known, so, the data (1.1.5), (1.1.6) are equivalent to (1.1.3), (1.1.4) (it suffices to set $G(t) = G_1(t) + hF(t)$ in (1.1.4)).

2) The point $x = 0$ is not the end of the string. The functions $F(t)$ and $G(t)$ involved in the boundary conditions (1.1.3), (1.1.4) are the result of measurement. Typical is the following situation: the string is homogeneous for $x < 0$ and the coefficients ρ and T are known constants. Equation (1.1.1) has the form $u_{tt} = c^2 u_{xx}$ for $x < 0$, where c is a constant. In this case the solution of the string equation is representable as

$$u|_{x<0} = f_1(t - x/c) + f_2(t + x/c).$$

The first term $f_1(t - x/c)$ models the incident wave (we assume that the string is immobile for $t < 0$, i. e., $f_1 = f_2 = 0$). The second term $f_2(t + x/c)$ is the wave which reflects from the string nonhomogeneities at points $x > 0$. If both the functions f_1 and f_2 are given, then, evidently,

$$u|_{x=0} = f_1(t) + f_2(t), \qquad u_x|_{x=0} = \frac{1}{c}(-\dot{f}_1(t) + \dot{f}_2(t))$$

(the point denotes the differentiation). It is clear that the data are reduced again to the case of (1.1.3), (1.1.4) where

$$F(t) = f_1(t) + f_2(t), \qquad G(t) = \frac{T}{c}(-\dot{f}_1(t) + \dot{f}_2(t)).$$

From condition (1.1.2), for $x > 0$, it follows that $F(t) = G(t) = 0$ for $t < 0$.

3) We may generalize the situation described above assuming the string to be inhomogeneous for $x < 0$ ($u|_{x<0}$ satisfies equation (1.1.1) as well). Two cases are possible, namely:

a) Properties of the string for $x < 0$ are also unknown, and the pair of the functions $F(t)$ and $G(t)$ is given. Then, there arise, in essence, the two inverse problems: find the coefficients of equation (1.1.1) for $x > 0$ and for $x < 0$. The second problem is reduced to the first if we make the change $x \to -x$. Note that the condition $u|_{t<0} = 0$ for $x > 0$ and for $x < 0$ denotes that the waves are generated by a point source located at the point $x = 0$.

b) Properties of the string are known for $x < 0$. The functions which describe the wave sources for $x < 0$ are also known. Then it suffices to set either $F(t)$ or $G(t)$. If we know one of these functions, we can solve the well-posed direct problem for $x < 0$ to recover $u(x,t)|_{x<0}$; from $u(x,t)$ given for $x = 0$, we find the second function ($G(t)$ or $F(t)$). Note that in this situation the string interval lying in the domain $x < 0$ may be both infinite and finite (in the latter case, we must impose a boundary condition of classical type on the left end of the string).

1.1.4. Reformulation of the problem in terms of a hyperbolic system

We make two changes of variables. The first, the change of the independent variable x, is of great importance. The second change (we pass from an equation of second order to a system of first order) is not so much important as convenient.

So, we introduce a new coordinate y instead of x:

$$y = \int_0^x \frac{dx}{c(x)}. \tag{1.1.7}$$

The physical sense of y is evident: it is the time taken for the wave to run along the string from zero to the point x. After this change the wave speed in the string is equal to unity. Since

$$\frac{\partial}{\partial x} = \frac{1}{c}\frac{\partial}{\partial y} = \sqrt{\frac{\rho}{T}}\frac{\partial}{\partial y},$$

equation (1.1.1) takes the form

$$u_{tt} = \frac{1}{\rho}\sqrt{\frac{\rho}{T}}\frac{\partial}{\partial y}T\sqrt{\frac{\rho}{T}}\frac{\partial u}{\partial y}$$

or

$$u_{tt} = \frac{1}{\sigma}\frac{\partial}{\partial y}\Big(\sigma\frac{\partial u}{\partial y}\Big). \tag{1.1.8}$$

The boundary condition (1.1.4) becomes as follows

$$\sigma\frac{\partial u}{\partial y}\Big|_{y=0} = G(t). \tag{1.1.9}$$

The second change concerns the unknown function. Equation (1.1.8) can be rewritten in the form

$$u_{tt} = u_{yy} + \frac{\sigma'_y}{\sigma}u_y.$$

We set $\tilde{\Psi}_1 = u_t - u_y$, $\tilde{\Psi}_2 = u_t + u_y$. Then

$$\frac{\partial\tilde{\Psi}_1}{\partial t} + \frac{\partial\tilde{\Psi}_1}{\partial y} = \frac{\partial\tilde{\Psi}_2}{\partial t} - \frac{\partial\tilde{\Psi}_2}{\partial y} = \frac{\sigma'_y}{\sigma}u_y = \frac{1}{2}\frac{\sigma'_y}{\sigma}(\tilde{\Psi}_2 - \tilde{\Psi}_1)$$

or

$$\frac{\partial\vec{\tilde{\Psi}}}{\partial t} + \Lambda\frac{\partial\vec{\tilde{\Psi}}}{\partial y} = \tilde{Q}\vec{\tilde{\Psi}}. \tag{1.1.10}$$

Here $\vec{\tilde{\Psi}}$ is the vector $(\tilde{\Psi}_1, \tilde{\Psi}_2)$, Λ and $\tilde{Q}$ are the following 2×2 matrices:

$$\Lambda = \begin{pmatrix} 1 & 0 \\ 0 & -1 \end{pmatrix}, \qquad \tilde{Q} = q \begin{pmatrix} -1 & 1 \\ -1 & 1 \end{pmatrix}, \qquad q = \frac{\sigma'}{2\sigma}.$$

Introduce a vector function $\vec{\Psi}$ by the formula $\vec{\tilde{\Psi}} = K\vec{\Psi}$, where K is a diagonal nondegenerate matrix. If K is chosen arbitrarily, the equation for $\vec{\Psi}$ keeps the form (1.1.10) with the only difference that $\tilde{Q}$ is replaced by a new matrix Q. Really, substituting $\vec{\tilde{\Psi}}$ by its expression, we obtain

$$K\frac{\partial \vec{\Psi}}{\partial t} + \Lambda K \frac{\partial \vec{\Psi}}{\partial y} + \Lambda K_y \vec{\Psi} = \tilde{Q} K \vec{\Psi}.$$

Hence, multiplying by K^{-1} and taking into account that Λ and K are commuting, we obtain

$$\frac{\partial \vec{\Psi}}{\partial t} + \Lambda \frac{\partial \vec{\Psi}}{\partial y} = (K^{-1}\tilde{Q}K - K^{-1}\Lambda K_y')\vec{\Psi} \overset{\text{def}}{=} Q\vec{\Psi}.$$

The matrix K can be chosen so that the diagonal elements of the matrix Q vanish. To this end, it suffices to set

$$K = \frac{1}{\sqrt{\sigma}} \begin{pmatrix} 1 & 0 \\ 0 & 1 \end{pmatrix}.$$

Then

$$Q = q \begin{pmatrix} 0 & 1 \\ -1 & 0 \end{pmatrix}. \tag{1.1.11}$$

We write once more the equation, the initial conditions and data of the inverse problem in terms of the function $\vec{\Psi}(y,t)$:

$$\frac{\partial \vec{\Psi}}{\partial t} + \Lambda \frac{\partial \vec{\Psi}}{\partial y} = Q\vec{\Psi}, \tag{1.1.12}$$

$$\vec{\Psi}|_{t<0} = 0. \tag{1.1.13}$$

Conditions (1.1.3), (1.1.4) are rewritten as follows

$$\vec{\Psi}|_{y=0} = \vec{H}(t) \tag{1.1.14}$$

where $\vec{H}(t) = (H_1(t), H_2(t)) = \sqrt{\sigma(0)}(\dot{F} - G, \dot{F} + G)$.

Remark 1.1.1. The function $\Psi_1(y,t)$ ($\Psi_2(y,t)$) may be physically interpreted as a wave which runs to the right (to the left). Really, in this case, when the string is homogeneous, i. e., $\sigma' = 0$, we obtain from system (1.1.12) that

$$\Psi_1(y,t) = \Psi_1(t-y) \qquad (\Psi_2(y,t) = \Psi_2(t+y)).$$

When $\sigma' \neq 0$, the running waves interact with medium nonhomogeneities and generate reflected waves which run in the opposite directions. Indeed, if we denote the right-hand side of system (1.1.12) as $Q\vec{\Psi} \equiv \vec{\Theta}$, then the components $\theta_1(y,t)$ and $\theta_2(y,t)$ of the vector function $\vec{\Theta}$ model the sources of waves which run to the right and to the left, respectively. The fact that

$$\theta_1(y,t) = \frac{\sigma'}{2\sigma}\Psi_2(y,t)$$

means that the functions modelling the source for the waves running to the right are proportional to those for the waves running to the left.

1.1.5. Determination of the string parameters from $\sigma(y)$ and additional information

From formulas (1.1.8), (1.1.9) of the previous section it follows that, given $\sigma(y)$ and $G(t)$, the function $u(y,t)$ and hence $F(t)$ are determined uniquely. Therefore, if we solve the inverse problem assuming $G(t)$ and $F(t)$ as known, we cannot obtain more information than contained in the function $\sigma(y)$. In order to find all the coefficients of equation (1.1.1) as functions of x (the distance from the point to the end of the string), we need an additional information[2]. Assume that we know one of the four functions $c(x)$, $\rho(x)$, $T(x)$, and $\sigma_0(x)$. Suppose that having solved the inverse problem, we find $\sigma(y)$.

In the fist case (when $c(x)$ is given), we can immediately find the function $y(x)$ by formula (1.1.7). Substituting $y(x)$, we find σ_0: $\sigma_0(x) = \sigma(y(x))$. If we know $\sigma_0(x)$ and $c(x)$, we find $\rho(x)$ and $T(x)$.

In the second and third cases (we know $\rho(x)$ or $T(x)$), we note that

$$y'_x = \frac{1}{c} = \frac{\rho}{\sigma} = \frac{\sigma}{T}. \tag{1.1.15}$$

If $\sigma(y)$ is known, and either $\rho(x)$ or $T(x)$ is known, then (1.1.15) is a differential equation of first order with separating variables relative to $y(x)$:

[2]The function σ as a function of the variable x should be distinguished from that of y. For convenience, we shall denote $\sigma(x)$ by $\sigma_0(x)$ in this section. So $\sigma_0(x) = \sigma(y(x))$.

$y'(x) = \rho(x)/\sigma(y)$ or $y'(x) = \sigma(y)/T(x)$. Solving this equation, we find

$$\int_0^y \sigma(y)\,\mathrm{d}y = \int_0^x \rho(x)\,\mathrm{d}x \qquad \text{or} \qquad \int_0^y \frac{\mathrm{d}y}{\sigma(y)} = \int_0^x \frac{\mathrm{d}x}{T(x)},$$

which gives the implicit expression of y in terms of x. If we know $\sigma_0(x) = \sigma(y(x))$ and $\rho(x)$ (or $T(x)$), we find the two remaining functions easily.

A more complicated situation is in the fourth case when the function $\sigma_0(x)$ is given and the function $\sigma(y)$ is found. We assume that $\sigma_0(x)$ is a continuous function which has a locally finite number of maxima and minima. Then the function σ has the same properties as $\sigma_0(x)$, namely, it has the same number of monotonicity intervals alternating in the same order and the same values at the extreme points. Otherwise, the information constituted in the problem data (*1)* in the function $\sigma_0(x)$ and *2)* in the pair of functions $F(t)$ and $G(t)$ which determine $\sigma(y)$) is inconsistent, and the problem is unsolvable.

If the indicated agreement of the properties of $\sigma(y)$ and $\sigma_0(x)$ holds, then we find $y(x)$ as follows: 1) we determine the interval of monotonicity of $\sigma_0(x)$ which contains x and find the corresponding monotonicity interval of $\sigma(y)$; 2) we find the solution $y(x)$ of equation $\sigma(y) = \sigma_0(x)$ relative to y (x is a parameter) which belongs to the found monotonicity interval (see Fig. 1.1).

Having determined $y(x)$, we easily find $c(x) = (y'_x)^{-1}$ and then the functions $\rho(x)$ and $T(x)$.

1.2. PECULIARITIES OF SOLUTION. FORMULATION OF THE DIRECT PROBLEM

1.2.1. Correct formulation of the direct problem

Further we shall consider the case when the source which generates the waves in the string is such that the corresponding solution of system (1.1.12) has the form

$$\Psi(y,t) = \delta(t-y)\begin{pmatrix}1\\0\end{pmatrix} + \begin{pmatrix}\psi_1(y,t)\\ \psi_2(y,t)\end{pmatrix}. \tag{1.2.1}$$

Here δ is the Dirac delta function, ψ_i ($i = 1,2$) are continuous functions for $t > y$ which may have discontinuities for $t = y$, $\operatorname{supp}\psi_i(y,t) \subseteq \{y,t : t \geq y\}$. As will be seen in the following, for the solution to be representable as (1.2.1), it is sufficient that $\sigma(y)$ be continuously differentiable.

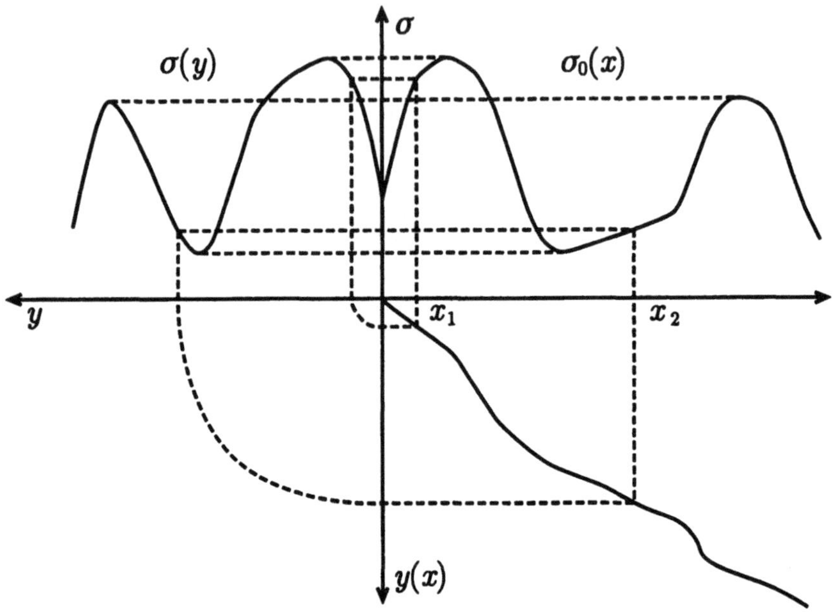

Fig. 1.1

Let us clarify what is meant by the solution of the direct problem. Suppose the components $\Psi_1(y,t)$, $\Psi_2(y,t)$ are related for $y=0$ through

$$\Psi_1(0,t) = \alpha\Psi_2(0,t) + M(t). \tag{1.2.2}$$

Here $\alpha = \text{const}$; $M(t)$ is a given function; $M(t) = \delta(t) + m(t)$. We shall do now some formal steps. The solution of the direct problem will be sought in the form (1.2.1). We define $\Psi_i(y,t)|_{t<y} \equiv 0$ and denote the independent variables by η and τ instead of y and t. Integrate the first equation of system (1.1.12), $\Psi_{1\tau} + \Psi_{1\eta} = q(\eta)\Psi_2(\eta,\tau)$, along the characteristic $\tau - \eta = t - y$ passing through a fixed point y, t:

$$\Psi_1(y,t) = \int_{t-y}^{t} q(\tau - t + y)\Psi_2(\tau - t + y, \tau)\,\mathrm{d}\tau + \Psi_1(0, t-y).$$

The second equation $\Psi_{2\tau} - \Psi_{2\eta} = -q(\eta)\Psi_1(\eta,\tau)$ is integrated along the characteristic $\tau + \eta = t + y$

$$\Psi_2(y,t) = -\int_0^t q(t - \tau + y)\Psi_1(t - \tau + y, \tau)\,\mathrm{d}\tau$$

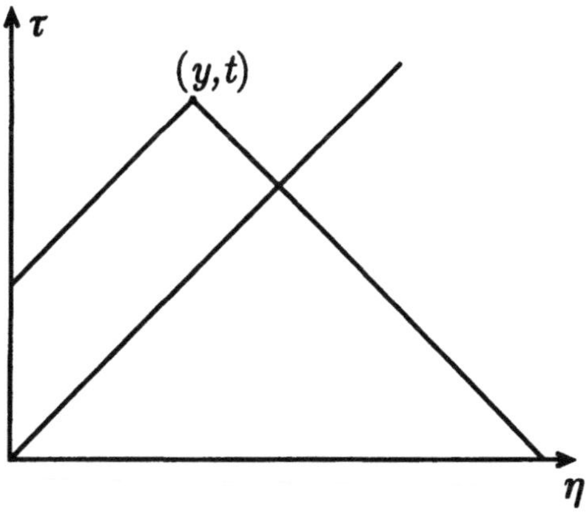

Fig. 1.2

considering that $\Psi_2(y,0) = 0$. The characteristics along which we integrate are illustrated in Fig. 1.2.

Taking into account that 1) $\vec{\Psi}|_{t<y} = 0$, 2) Ψ_1 has the δ-singularity for $t = y$, 3) $\Psi_i(y,t) = \psi_i(y,t)$ for $t > y$, we obtain for $t > y$

$$\psi_1(y,t) = \int_{t-y}^{t} q(\tau - t + y)\psi_2(\tau - t + y, \tau)\,\mathrm{d}\tau + \psi_1(0, t-y), \tag{1.2.3}$$

$$\psi_2(y,t) = -\int_{(t+y)/2}^{t} q(t + y - \tau)\psi_1(t + y - \tau, \tau)\,\mathrm{d}\tau - \frac{1}{2} q\Big(\frac{t+y}{2}\Big). \tag{1.2.4}$$

The last term in (1.2.4) is the contribution of the δ-term into the integral. In (1.2.3) we replace $\psi_1(0, t-y)$ by its expression from the boundary condition (1.2.2). In view of (1.2.4), we obtain

$$\begin{aligned}\psi_1(y,t) = &\int_{t-y}^{t} q(\tau - t + y)\psi_2(\tau - t + y, \tau)\,\mathrm{d}\tau\\ &- \alpha\Big(\int_{(t-y)/2}^{t-y} q(t - y - \tau)\psi_1(t - y - \tau, \tau)\,\mathrm{d}\tau + \frac{1}{2} q\Big(\frac{t-y}{2}\Big)\Big)\\ &+ m(t-y).\end{aligned} \tag{1.2.5}$$

Relations (1.2.4), (1.2.5), for known q, m, α, represent a system of Volterra integral equations of second kind. It is known that such systems are solvable. Solutions of such systems are continuous functions in y, t.

Definition 1.2.1. A generalized function of t which involves y as a parameter and vanishes for $t < y$ is called a solution of the problem for equation (1.1.12) subject to the zero initial condition and boundary condition (1.2.2) if it is representable in the form (1.2.1), with ψ_1, ψ_2 being continuous for $t > y$ and satisfying (1.2.4), (1.2.5).

Remark 1.2.1. It is easy to verify that the boundary condition (1.2.2) is equivalent, for $\alpha = 1$, to the Neyman condition for the string equation

$$Tv_x|_{x=0} = -\frac{1}{2\sqrt{\sigma(0)}} M(t)$$

and, for $\alpha = -1$, to the Dirichlet condition

$$v|_{x=0} = \frac{1}{2\sqrt{\sigma(0)}} \int_0^t M(t)\, dt.$$

Condition (1.2.2) for $\alpha = 0$ corresponds to the direct scattering problem for the string which is homogeneous for $x < 0$:

$$u|_{t<0} = f_1(t - x/c) \qquad (f_1|_{t<x/c} \equiv 0), \qquad M(t) = 2\sqrt{\sigma(0)} = f_1'(t).$$

In the case of elastic fixing of the string, the boundary condition

$$(Tu_x - hu)|_{x=0} = G_1(t)$$

is equivalent (in terms of system (1.1.12)) to the condition

$$\psi_2(0,t) - \psi_1(0,t) = \frac{h}{\sigma} \int_0^t (\psi_1(0,t) + \psi_2(0,t))\, dt + \frac{1}{\sigma(0)} G_1(t).$$

This formula, after simple transformations, is reduced to

$$\psi_1(0,t) = \psi_2(0,t) - \frac{2h}{\sigma(0)} \int_0^t \exp\Big(\frac{h}{\sigma}(t-\tau)\Big) \psi_2(0,\tau)\, d\tau - \frac{1}{\sigma(0)} G_1(t)$$
$$+ h\sigma(0)^{-3/2} \int_0^t \exp\Big(\frac{h}{\sigma}(t-\tau)\Big) G_1(\tau)\, d\tau.$$

We suppose here that $G_1(t)$ has the following form

$$G_1(t) = -\sigma(0)\delta(t) + g_1(t), \qquad g_1(t)|_{t<0} = 0, \qquad g_1(t)|_{t\geq 0} \in C.$$

In this case the system of integral equations (1.2.4), (1.2.5) of the direct problem should be modified. Equation (1.2.4) retains its form. To obtain the analog of equation (1.2.5), we should first replace $\psi_1(0, t-y)$ in (1.2.3) by its expression from the last boundary condition and then get rid of $\psi_2(0,t)$ and $\psi_2(0,\tau)$ using (1.2.4).

1.2.2. Singularities of solution of the hyperbolic system

The subsequent discussion may be led on the basis of the above rigorous setting of the direct problem. However, we prefer to use a nonstrict setting in the differential form. This allows us to make our presentation more clear.

If $\sigma(y)$ and $m(t)$ are sufficiently smooth, the solution of system (1.1.12) satisfying the boundary condition (1.2.2) will be as follows

$$\vec{\Psi}(y,t) = \begin{pmatrix} a\delta(t-y) \\ 0 \end{pmatrix} + \begin{pmatrix} b_1 \\ b_2 \end{pmatrix} \varepsilon(t-y) + \begin{pmatrix} c_1 \\ c_2 \end{pmatrix} (t-y)_+ + \dots . \quad (1.2.6)$$

Hereafter we define

$$s_+^\lambda = \begin{cases} s^\lambda, & \text{for } s > 0, \\ 0, & \text{for } s \le 0, \end{cases} \qquad \varepsilon(s) = s_+^0, \qquad s_+ = s_+^1.$$

The coefficients a, b_1, b_2, c_1, c_2 are functions dependent on y; smoother terms for $t = y$ than written down in (1.2.6) are denoted by dots. Let us find a, b_1, b_2, c_1, c_2. To this purpose, we substitute (1.2.6) into (1.1.12) and equate the coefficients at similar singularities. We take into account that for each function χ

$$\Big(\frac{\partial}{\partial t} + \frac{\partial}{\partial y}\Big)\chi(t-y) = 0,$$

$$\Big(\frac{\partial}{\partial t} - \frac{\partial}{\partial y}\Big)\varepsilon(t-y) = 2\delta(t-y), \qquad \Big(\frac{\partial}{\partial t} - \frac{\partial}{\partial y}\Big)(t-y)_+ = 2\varepsilon(t-y).$$

As a result, we obtain

$$a'\begin{pmatrix} \delta(t-y) \\ 0 \end{pmatrix} + 2\begin{pmatrix} 0 \\ b_2 \end{pmatrix}\delta(t-y) + \begin{pmatrix} b_1' \\ -b_2' \end{pmatrix}\varepsilon(t-y)$$

$$+ 2\begin{pmatrix} 0 \\ c_2 \end{pmatrix}\varepsilon(t-y) + \begin{pmatrix} c_1' \\ -c_2' \end{pmatrix}(t-y)_+ + \dots$$

$$= \frac{\sigma'}{2\sigma}\begin{pmatrix} 0 \\ -a \end{pmatrix}\delta(t-y) + \frac{\sigma'}{2\sigma}\begin{pmatrix} b_2 \\ -b_1 \end{pmatrix}\varepsilon(t-y) + \frac{\sigma'}{2\sigma}\begin{pmatrix} c_2 \\ -c_1 \end{pmatrix}(t-y)_+ + \dots .$$

Equating the terms with the same singularities, we obtain $a' = 0$, whence,

$$a = a_0 = \text{const}, \qquad b_2 = -a_0\frac{\sigma'}{4\sigma}, \qquad b_1' = b_2\frac{\sigma'}{4\sigma} = \frac{-a_0}{8}\Big(\frac{\sigma'}{\sigma}\Big)^2.$$

Consequently,

$$b_1 = \frac{-a_0}{8} \int_0^y \left(\frac{\sigma'}{\sigma}\right)^2 dy + b_0 \qquad (b_0 = \text{const}).$$

Evidently, we can analogously obtain an arbitrary number of terms of the $\vec{\Psi}$ expansion. More precisely, we can obtain so many terms as smoothness of σ admits. Note that the requirement on the smoothness of σ becomes stronger with increasing number of terms of the expansion. For example,

$$c_2 = -\frac{\sigma'}{4\sigma} b_1 + \frac{1}{2} b_2' = -\frac{\sigma'}{4\sigma} b_1 - \frac{a_0}{8} \left(\frac{\sigma'}{\sigma}\right)'$$

depends on the second derivative of σ. The same tendency holds for the succeeding terms.

To find the constants a_0 and b_0, we substitute (1.2.6) into the boundary condition (1.2.2):

$$\begin{aligned} a_0 \delta(t) + (b_1 - \alpha b_2)|_{y=0} \varepsilon(t) + (c_1 - \alpha c_2)|_{y=0} t_+ + \dots \\ = \delta(t) + m(0)\varepsilon(t) + m'(0) t_+ + \dots . \end{aligned}$$

Hence,

$$a_0 = 1, \quad (b_1 - \alpha b_2)|_{y=0} = b_0 + \alpha \frac{\sigma'(0)}{4\sigma(0)} = m(0), \quad b_0 = -\alpha \frac{4\sigma'(0)}{\sigma(0)} + m(0).$$

So, the first two terms of the expansion of the vector function $\vec{\Psi}(y,t)$ are as follows:

$$\vec{\Psi}(y,t) = \begin{pmatrix} 1 \\ 0 \end{pmatrix} \delta(t-y) + \begin{pmatrix} -\alpha \dfrac{\sigma'(0)}{4\sigma(0)} + m(0) - \dfrac{1}{8} \displaystyle\int_0^y \left(\frac{\sigma'}{\sigma}\right)^2 dy \\ -\dfrac{1}{4} \dfrac{\sigma'(y)}{\sigma(y)} \end{pmatrix} \varepsilon(t-y) + \dots . \tag{1.2.7}$$

In particular, (1.2.7) yields the relation

$$\Psi_2(y, y+0) = -\frac{1}{4} \frac{\sigma'(y)}{\sigma(y)}. \tag{1.2.8}$$

1.2.3. Singularities of solution of the string equation

An analog of the above expansion can be obtained for the solution of the problem (1.1.1), (1.1.2), (1.1.4). We shall give it for the equation in the more general form

$$\rho u_{tt} = \frac{\partial}{\partial x}(Tu_x) - q(x)u \tag{1.2.9}$$

with the initial condition (1.1.2) and the boundary condition

$$(Tu_x - hu)|_{x=0} = G_1(t). \tag{1.2.10}$$

We suppose that $G_1(t)$ has the following expansion

$$G_1(t) = \delta(t) + g_1(0)\varepsilon(t) + \dots . \tag{1.2.11}$$

We seek for the solution in the form

$$u(x,t) = \alpha(x)\varepsilon(t-\tau(x)) + \beta(x)(t-\tau(x))_+ + \frac{\gamma(x)}{2}(t-\tau(x))_+^2 + \dots .$$

Substituting this expansion into (1.2.9), we obtain

$$\begin{aligned}
&\rho\alpha(x)\delta'(t-\tau) + \rho\beta(x)\delta(t-\tau) + \rho\gamma(x)\varepsilon(t-\tau) + \dots \\
&\quad = \frac{\partial}{\partial x}\Big(-T\alpha\tau_x'\delta(t-\tau) + T\alpha_x'\varepsilon(t-\tau) - T\beta\tau_x'\varepsilon(t-\tau) \\
&\qquad + T\beta_x'(t-\tau)_+ - T\gamma\tau_x'(t-\tau)_+ + \dots\Big) - q\alpha\varepsilon(t-\tau) + \dots \\
&\quad = T\alpha(\tau_x')^2\delta'(t-\tau) - (T\alpha\tau_x')_x'\delta(t-\tau) - T\alpha_x'\tau_x'\delta(t-\tau) \\
&\qquad + (T\alpha_x')_x'\varepsilon(t-\tau) + T\beta(\tau_x')^2\delta(t-\tau) - (T\beta\tau_x')_x'\varepsilon(t-\tau) \\
&\qquad - T\beta_x'\tau_x'\varepsilon(t-\tau) + T\gamma(\tau_x')^2\varepsilon(t-\tau) - q\alpha\varepsilon(t-\tau) + \dots .
\end{aligned}$$

Equating the coefficients at $\delta'(t-\tau)$, we find that $(\tau')^2 = \rho/T$, whence

$$\tau = \int_0^x \sqrt{\frac{\rho(\xi)}{T(\xi)}}\,\mathrm{d}\xi = \int_0^x \frac{\mathrm{d}\xi}{c(\xi)}.$$

Equating the coefficients at $\delta(t-\tau)$, we obtain

$$-(T\alpha\tau_x')_x' - T\alpha_x'\tau_x' = 0.$$

Taking into account the obtained value of τ_x' and the fact that $\sqrt{\rho T} = \sigma$, we have

$$(\alpha\sigma)_x' + \alpha_x'\sigma = 0.$$

This relation can be rewritten as

$$2\sigma^{1/2}(\alpha\sigma^{1/2})'_x = 0,$$

whence

$$\alpha(x) = \sigma^{-1/2} \cdot \text{const}\,.$$

The boundary condition (1.1.4) and expression (1.2.1) yield

$$\alpha(x) = -\frac{1}{\sqrt{\sigma(x)\sigma(0)}}\,. \tag{1.2.12}$$

Finally, equating the coefficients at $\varepsilon(t-\tau)$, we obtain

$$(T\alpha'_x)'_x - (T\beta\tau'_x)'_x - T\beta'_x\tau'_x - \alpha q(x) = 0$$

or

$$(\sigma\beta)'_x + \sigma\beta'_x = 2\sigma^{1/2}(\sigma^{1/2}\beta)'_x = (T\alpha'_x)'_x - \alpha q.$$

Taking into account the expression for $\alpha(x)$ and boundary condition (1.2.10), we have

$$\beta(x) = -\frac{1}{2\sqrt{\sigma(0)\sigma(x)}}\int_0^x \sigma^{-1/2}(\xi)\left[(T(\xi)(\sigma^{-1/2})'_\xi)'_\xi + \sigma^{-1/2}q\right]\mathrm{d}\xi + \frac{\text{const}}{\sqrt{\sigma(x)}}$$

$$= \frac{1}{2\sqrt{\sigma(0)\sigma(x)}}\int_0^x \left[T(\xi)((\sigma^{-1/2})'_\xi)^2 - \frac{q(\xi)}{\sigma(\xi)}\right]\mathrm{d}\xi - \frac{T(x)}{2\sqrt{\sigma(0)\sigma(x)}}(\sigma^{-1/2})'_x$$

$$+ \frac{1}{\sqrt{\sigma(x)\sigma(0)}}\left(\frac{\sigma'_x(0)}{\rho(0)} + \frac{h}{\sigma(0)} - g_1(0)\right) + \frac{T(0)}{2\sigma(0)\sqrt{\sigma(x)}}\,(\sigma^{-1/2})'_x|_{x=0}.$$

In particular, when $\sigma(x) \equiv 1$, the expression for $\beta(x)$ gives

$$u_t|_{t=\tau(x)+0} = -\frac{1}{2}\int_0^x q(\xi)\,\mathrm{d}\xi - g_1(0) + h. \tag{1.2.13}$$

1.2.4. Singularities of solution in case of discontinuous coefficients

As noted above, the obtained formulas hold for sufficiently smooth coefficients ρ and T of equation (1.1.1). If the properties of the string are changed so that the coefficients lose their continuity or smoothness, then reflected waves arise at the points of discontinuity. This leads to a change of peculiarities of the wave field.

We consider the case when the functions $\rho(x)$ and $T(x)$ are discontinuous at a point x_0. We assume also that the function u is continuous at the point x_0, and so is Tu'_x. In terms of the coordinate $y = \int_0^x (1/c(\xi))\, d\xi$ the last condition means the continuity of $\sigma u'_y$. Passing, as above, to the vector function

$$\vec{\Psi}(y,t) = (\sqrt{\sigma}(u_t - u_y), \sqrt{\sigma}(u_t + u_y)),$$

we note that the indicated conditions of sewing together for the function u are equivalent to the following conditions for $\vec{\Psi}$: the functions $(\Psi_1+\Psi_2)/\sqrt{\sigma}$ and $(\Psi_2-\Psi_1)/\sqrt{\sigma}$ must be continuous. If we denote by $\vec{\Psi}^{\pm} = (\Psi_1^{\pm}, \Psi_2^{\pm})$ the limit values of $\vec{\Psi}$ for $y \to y_0 \pm 0$ ($y_0 = y(x_0)$) and by $\sigma^{\pm}$ the corresponding limit values of $\sigma(y)$, then these sewing conditions mean that

$$(\Psi_1^+ + \Psi_2^+)(\sigma^+)^{-1/2} = (\Psi_1^- + \Psi_2^-)(\sigma^-)^{-1/2},$$

$$(\Psi_1^+ - \Psi_2^+)(\sigma^+)^{-1/2} = (\Psi_1^- - \Psi_2^-)(\sigma^-)^{-1/2}.$$

Deriving Ψ_1^+, Ψ_2^+ from these formulas and denoting $\sigma^+/\sigma^- = \exp(2\nu)$, we obtain

$$\vec{\Psi}^+ = S(\nu)\vec{\Psi}^- \tag{1.2.14}$$

where $S(\nu)$ is the following matrix:

$$S(\nu) = \begin{pmatrix} \cosh\nu & \sinh\nu \\ \sinh\nu & \cosh\nu \end{pmatrix}. \tag{1.2.15}$$

The transformation governed by the above matrix is called the hyperbolic rotation (the Lorentz transformation). The expression

$$(\Psi_1^-)^2 - (\Psi_2^-)^2 = (\Psi_1^+)^2 - (\Psi_2^+)^2$$

is invariant under this transformation. The matrices $S(\nu)$ form the one-parameter group

$$S(\nu_1)S(\nu_2) = S(\nu_1 + \nu_2).$$

In particular, $S(-\nu) = S(\nu)^{-1}$, i.e.,

$$\vec{\Psi}^- = S(-\nu)\vec{\Psi}^+. \tag{1.2.16}$$

Assume that the wave $\Psi_1(y,t)$ incident from the left has the singularity

$$\Psi_1 = \delta(t-y) + \dots, \qquad y < y_0.$$

After interaction of this wave with the jump of the function σ, two waves will arise: the past wave

$$\Psi_1 = s\delta(t-y) + \dots, \qquad y > y_0,$$

and the reflected wave

$$\Psi_2 = r\delta(t+y-2y_0) + \dots, \qquad y < y_0.$$

As follows from the results of Subsection 1.2.1, the coefficients s and r are constant: s is called the passage coefficient, while r is referred to as the reflection coefficient. It is easy to see that relations (1.2.14), (1.2.16) hold not only for $\vec{\Psi}^{\pm}$ but also for the coefficients of the δ-terms. Hence

$$\begin{pmatrix} 1 \\ r \end{pmatrix} = \begin{pmatrix} \cosh\nu & -\sinh\nu \\ -\sinh\nu & \cosh\nu \end{pmatrix} \begin{pmatrix} s \\ 0 \end{pmatrix},$$

i. e., $1 = s\cosh\nu$, $r = -s\sinh\nu$, therefore

$$s = 1/\cosh\nu, \qquad r = -\tanh\nu. \tag{1.2.17}$$

Let $y_0 > 0$. Consider the reflected wave for $y = 0$. Suppose that the incident wave has the singularity $\delta(t-y)$. Then in the reflected wave for $y = 0$ there will appear a term with the singularity $r\delta(t-2y_0) + \dots$. If we know the moment $2y_0$ of arrival of this singularity, we can find the coordinate y_0 of this jump. If we know the coefficient r ($|r| < 1$), we can find σ^+/σ^-. Really,

$$r = -\tanh\nu = -\frac{e^{\nu} - e^{-\nu}}{e^{\nu} + e^{-\nu}} = -\frac{e^{2\nu} - 1}{e^{2\nu} + 1} = -\frac{\sigma^+/\sigma^- - 1}{\sigma^+/\sigma^- + 1},$$

whence

$$\frac{\sigma^+}{\sigma^-} = \frac{1-r}{1+r}. \tag{1.2.18}$$

Thus, if we know the singularities of the reflected wave we determine the point of the jump and the ratio σ^+/σ^- which characterizes the value of the jump.

Of special note is the scattering problem in the case when the point of the jump is at zero. We know the incident and reflected waves for $y = -0$ or, which is equivalent, the whole wave field in a left-sided neighborhood of zero. The indication of such situation is a δ-term in the reflected field with the singularity at zero if such a term is in the incident wave. From the coefficients of the δ-terms we can find σ^+/σ^-; therefore, we can calculate the limit values of the wave field for $y \to +0$. Thus, the inverse scattering problem, if we have a jump of $\sigma(y)$ at zero, is reduced to the problem (1.1.1)–(1.1.4). In this case, we take for $u|_{x=0}$ and $u_x|_{x=0}$ their limits as $x \to +0$.

1.3. THE FIRST METHOD OF SOLUTION OF INVERSE PROBLEM

1.3.1. Derivation of the system of integral equations

As was mentioned above, the method is based on reducing the problem to a system of nonlinear Volterra integral equations. We assume that $\sigma(y)$ is a continuous piecewise smooth function. Consider system (1.1.12). The lines $t = \pm y + t_0$ are called the characteristics of system (1.1.12)[3]. The expressions in the left-hand side of system (1.1.12) are the derivatives of Ψ_1 or Ψ_2 taken along the corresponding characteristic. If we substitute $t = y + t_0$ for t in the second argument of Ψ_1, then

$$\frac{\mathrm{d}}{\mathrm{d}y}\Psi_1(y, y + t_0) = \frac{\partial \Psi_1}{\partial y} + \frac{\partial \Psi_1}{\partial t}\bigg|_{t=y+t_0}.$$

Analogously,

$$\frac{\mathrm{d}}{\mathrm{d}y}\Psi_2(y, -y + t_0) = \frac{\partial \Psi_2}{\partial y} - \frac{\partial \Psi_2}{\partial t}\bigg|_{t=-y+t_0}.$$

Therefore, integrating the equations entering system (1.1.12) along the corresponding characteristic (see Fig. 1.3) we obtain for $t > y$

$$\Psi_1(y,t) = \int_0^y q(\eta)\Psi_2(\eta, t + \eta - y)\,\mathrm{d}\eta + \Psi_1(0, t - y),$$

$$\Psi_2(y,t) = \int_0^y q(\eta)\Psi_1(\eta, t - \eta + y)\,\mathrm{d}\eta + \Psi_2(0, t + y).$$

Here $q(y) = \sigma'/2\sigma$. Substituting the functions $\Psi_i(0, t \pm y)$ by their values $h_i(t \pm y)$, we have

$$\Psi_1(y,t) = \int_0^y q(\eta)\Psi_2(\eta, t + \eta - y)\,\mathrm{d}\eta + h_1(t - y), \tag{1.3.1}$$

---

[3]In the general case when the system has the form

$$\frac{\partial \vec{\Psi}}{\partial t} + \Lambda \vec{\Psi}'_y = Q\vec{\Psi}$$

where $\vec{\Psi} \in \mathbb{R}^n$, Λ and Q are $n \times n$ matrices, Q is an arbitrary matrix, Λ is a diagonal matrix with real diagonal elements $\lambda_i(y,t) \neq 0$, $i = 1, \dots, n$; by the characteristics are meant the curves $t = t_i(y)$ where t_i are the solutions of the differential equation $t'_{iy} = 1/\lambda_i(y,t)$. If $\lambda_i = \lambda_i(y)$ are independent of t, these differential equations are solved explicitly

$$t_i = \int_0^y \frac{\mathrm{d}\eta}{\lambda_i(\eta)} + t_0.$$

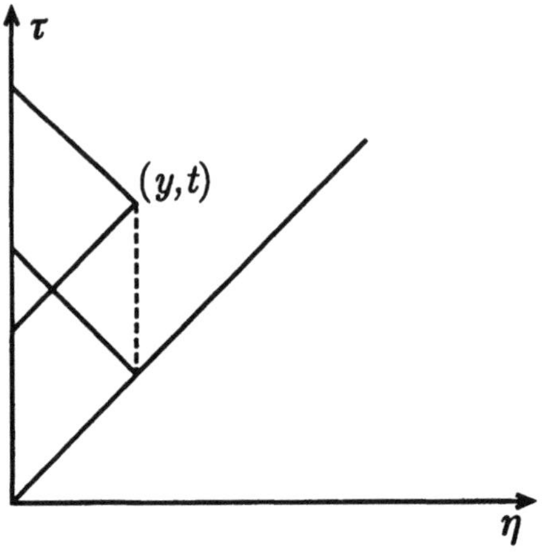

Fig. 1.3

$$\Psi_2(y,t) = \int_0^y q(\eta)\Psi_1(\eta, t-\eta+y)\,d\eta + h_2(t+y). \qquad (1.3.2)$$

Remark 1.3.1. The function $\vec{\Psi}(y,t)$ has a δ-singularity for $t = y$. However, this fact is not of importance because, for $t > y$, the characteristics passing through the point (y,t) (we integrate along these characteristics in (1.3.1) and (1.3.2)) do not intersect the δ-term.

We have the three unknown functions $\Psi_1(y,t)$, $\Psi_2(y,t)$, and $q(y)$ in equations (1.3.1) and (1.3.2). Therefore, we need one more equation. Recall that, if the function $G(t)$ has singularity (1.2.1), formula (1.2.8) holds, that is,

$$\Psi_2(y,y) \overset{\text{def}}{=} \Psi_2(y,y+0) = -\frac{1}{4}\frac{\sigma'(y)}{\sigma(y)} = -\frac{1}{2}q(y).$$

As a result we obtain

$$q(y) = -2\int_0^y q(\eta)\Psi_1(\eta, 2y-\eta)\,d\eta - 2h_2(2y). \qquad (1.3.3)$$

This is the desired equation. The triple of equations (1.3.1)–(1.3.3) forms the system of nonlinear Volterra integral equations of second kind (see Fig. 1.3).

1.3.2. Investigation of the system of integral equations

The subsequent examination of the system (1.3.1)–(1.3.3), while conventional, is rather awkward. Therefore, the reader who prefers not to go into

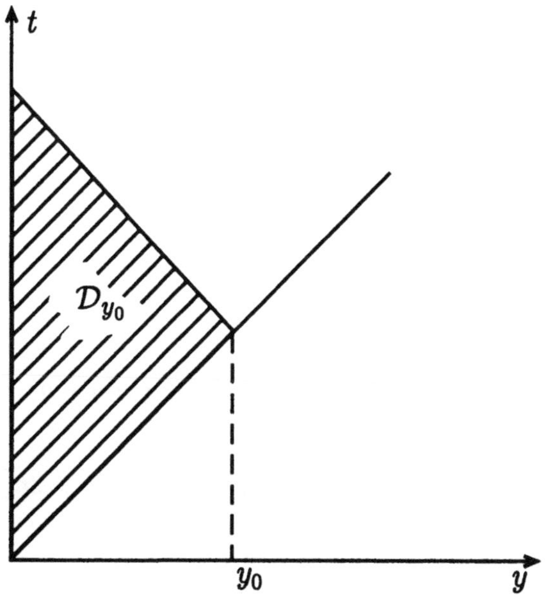

Fig. 1.4

an extended discussion of this subject may miss this and the next subsections without damage to the following study.

We shall show that the solution of this system exists, is unique and depends continuously on the problem data (the sense of the assertions will be clear further).

Consider the set of triples of functions

$$X = \{\Psi_1(y,t), \Psi_2(y,t), q(y)\}.$$

The functions which enter the triple X are also denoted by $X_1(y,t)$, $X_2(y,t)$, $X_3(y)$. The function $X_3(y) = q(y)$ defined on the interval $[0, y_0]$, for some $y_0 > 0$, is continuous; $X_i(y,t) = \Psi_i(y,t)$, $i = 1, 2$, defined on the triangle $\mathcal{D}_{y_0} : y \leq y_0,\ y \leq t \leq 2y_0 - y$, are continuous as well (see Fig 1.4).

Suppose that

$$\max(|\Psi_1(y,t)|, |\Psi_2(y,t)|, |q(y)|) = M$$

where $M > 0$ is a constant. The set of all the triples is denoted by B_{M,y_0}. Suppose $X^{(1)}$ and $X^{(2)}$ to be two such triples. We introduce the distance between them as follows

$$\rho(X^{(1)}, X^{(2)}) = \max_{y\in[0,y_0],\, y,t\in\mathcal{D}_{y_0}} \left\{|\Psi_1^{(1)}(y,t) - \Psi_1^{(2)}(y,t)|\,,\right.$$

$$\left.|\Psi_2^{(1)}(y,t) - \Psi_2^{(2)}(y,t)|, |q^{(1)}(y) - q^{(2)}(y)|\right\}.$$

We define an operator K which maps a triple X into a triple KX as follows:

$$(KX)_1(y,t) = \int_0^y q(\eta)\Psi_2(\eta, t+\eta-y)\,\mathrm{d}\eta + h_1(t-y),$$

$$(KX)_2(y,t) = \int_0^y q(\eta)\Psi_1(\eta, t-\eta+y)\,\mathrm{d}\eta + h_2(t+y),$$

$$(KX)_3(y) = -2\int_0^y q(y)\Psi_1(\eta, 2y-\eta)\,\mathrm{d}\eta - 2h_2(2y).$$

If the functions $h_1(t)$ and $h_2(t)$ are continuous for $0 \le t \le 2y_0$, then, evidently, the triple KX belongs to the set B_{M',y_0} with another constant M'. The system (1.3.1)–(1.3.3) can be written briefly as

$$X = KX. \tag{1.3.4}$$

Let $|h_i| \le m$ for $i = 1, 2$. It is easy to see that

$$|(KX)_i(y,t)| = \left|\int_0^y q(\eta)\Psi_{\bar{\imath}}(\eta, t \pm (\eta - y))\,\mathrm{d}\eta + h_i(t \mp y)\right| \le M^2 y_0 + m.$$

Here $i = 1, 2$; $\bar{\imath} = 2$ if $i = 1$, and $\bar{\imath} = 1$ if $i = 2$. An analogous estimate holds for $|(KX)_3|$:

$$|(KX)_3| \le 2M^2 y_0 + 2m.$$

Thus, if $2M^2y_0 + 2m \le M$ (this can be achieved for $M > 2m$ and y_0 sufficiently small), then the operator K maps the set B_{M,y_0} into itself.[4]

Suppose that $X^{(1)}$ and $X^{(2)}$ are two triples which belong to B_{M,y_0}. We assume that

$$|X_i^{(1)}(y,t) - X_i^{(2)}(y,t)| \le Cy^n, \qquad i = 1, 2,$$

$$|X_3^{(1)}(y) - X_3^{(2)}(y)| \le Cy^n \tag{1.3.5}$$

[4]We choose, for a fixed $M > 2m$, the maximal possible value of y_0 for which

$$KB_{M,y_0} \subset B_{M,y_0} \quad : \quad y_0 = \frac{M - 2m}{2M^2}.$$

Changing M, we find the maximum of such y_0. It is easy to verify that the maximal y_0 is achieved for $M = 4m$. In this case, $y_{0\,\max} = 1/16m$.

where $C > 0$ is a constant; n is integer, $n \geq 0$. Let us evaluate

$$\begin{aligned}
&|(KX^{(1)})_1 - (KX^{(2)})_1| \\
&\quad = \left| \int_0^y (X_3^{(1)}(\eta) X_2^{(1)}(\eta, t-y+\eta) - X_3^{(2)}(\eta) X_2^{(2)}(\eta, t-y+\eta)) \, d\eta \right| \\
&\quad \leq \int_0^y |(X_3^{(1)}(\eta) - X_3^{(2)}(\eta)) X_2^{(1)}(\eta, t-y+\eta)| \, d\eta \\
&\qquad + \int_0^y |X_3^{(2)}(\eta)(X_2^{(1)}(\eta, t-y+\eta) - X_2^{(2)}(\eta, t-y+\eta))| \, d\eta.
\end{aligned}$$

Estimating the differences $|X_3^{(1)}(\eta) - X_3^{(2)}(\eta)|$ and $|X_2^{(1)}(\eta, t-y+\eta) - X_2^{(2)}(\eta, t-y+\eta)|$ via $C\eta^n$ and the multipliers $X_2^{(1)}(\eta, t-y+\eta)$, $X_3^{(2)}(\eta)$ via M, we obtain

$$|(KX^{(1)})_1(y,t) - (KX^{(2)})_1(y,t)| \leq 2MC \int_0^y \eta^n \, d\eta \leq \frac{4}{n+1} MCy^{n+1}.$$

The differences $|(KX^{(1)})_i - (KX^{(2)})_i|$ for $i = 2, 3$ are estimated analogously. So, the triples $KX^{(1)}$, $KX^{(2)}$ satisfy the estimate of the type (1.3.5) with the exponent $n+1$ instead of n and with the constant $4MC/(n+1)$ instead of C. Suppose $X^{(1)}$ and $X^{(2)}$ to satisfy the estimates (1.3.5) with $n = 0$. Evidently, we may take $\rho(X^{(1)}, X^{(2)})$ as C. Then, $KX^{(1)}$ and $KX^{(2)}$ satisfy the estimates (1.3.5) with $n = 1$ and the constant $4MC/1$. Hence, it follows that $K^2X^{(1)}$ and $K^2X^{(2)}$ satisfy (1.3.5) with $n = 2$ and the constant $(4M)^2C/(1 \cdot 2)$; $K^3X^{(1)}$ and $K^3X^{(2)}$ satisfy (1.3.5) with $n = 3$ and the constant $(4M)^3C/(1 \cdot 2 \cdot 3)$. By induction it is easy to show that $K^nX^{(1)}$ and $K^nX^{(2)}$ satisfy the estimate (1.3.5) with the exponent n and the constant $(4M)^nC/n!$. Here $K^2X^{(i)}$, $K^3X^{(i)}$, ... results from double triple, etc., application of the operator K to $X^{(i)}$. Substituting y^n by its maximal value y_0^n and C by $\rho(X^{(1)}, X^{(2)})$, we obtain

$$\rho(K^nX^{(1)}, K^nX^{(2)}) \leq \frac{(4My_0)^n}{n!} \rho(X^{(1)}, X^{(2)}).$$

Evidently, $(4My_0)^n/n! \to 0$ as $n \to \infty$ for each value of My_0. Therefore, if $n = n_0$ is sufficiently large,

$$\frac{(4My_0)^{n_0}}{n_0!} \equiv \alpha < 1$$

and K^{n_0} is contractive. By the Banach contraction mapping principle (see Kolmogorov and Fomin (1981)) the equation

$$X = K^{n_0} X \tag{1.3.6}$$

has a solution. This solution is unique. Evidently, if X solves equation (1.3.4), it satisfies equation (1.3.6). Therefore, solution of equation (1.3.4) is unique as well. Suppose X to be a solution of equation (1.3.6). Applying the operator K to both the sides of equation (1.3.6), we obtain

$$KX = K^{n_0+1}X = K^{n_0}(KX).$$

Hence, KX solves equation (1.3.6). Since (1.3.6) has a unique solution, we have $KX = X$, i. e., X satisfies equation (1.3.4). Solution of equation (1.3.4) may be obtained by the successive approximations method.

1.3.3. Continuous dependence of solution on inverse problem data

The operator K defined above depends on the functions h_1, h_2 entering the inverse problem setting. Suppose that we have two pairs of the functions $h_1(t)$, $h_2(t)$ and $\tilde{h}_1(t)$, $\tilde{h}_2(t)$; the operators K and $\tilde{K}$ are constructed using the functions $h_i(t)$, $\tilde{h}_i(t)$, respectively. The difference $(KX) - (\tilde{K}X)$, for each X, is the triple

$$h_1(t-y) - \tilde{h}_1(t-y), \qquad h_2(t+y) - \tilde{h}_2(t+y), \qquad h_2(2y) - \tilde{h}_2(2y).$$

Evidently, the triple is independent of the choice of X. Let $\rho(KX, \tilde{K}X) \leq \delta$, and let X and $\tilde{X}$ be solutions of the equations $X = KX$ and $\tilde{X} = \tilde{K}\tilde{X}$. Let $\alpha = (4My_0)^{n_0}/n_0! < 1$. Then

$$\begin{aligned}\rho(X, \tilde{X}) &= \rho(K^{n_0}X, \tilde{K}^{n_0}\tilde{X}) \leq \rho(K^{n_0}X, \tilde{K}^{n_0}X) + \rho(\tilde{K}^{n_0}X, \tilde{K}^{n_0}\tilde{X}) \\ &\leq \rho(\tilde{K}^{n_0}X, K^{n_0}X) + \alpha\rho(X, \tilde{X}),\end{aligned}$$

whence

$$\rho(X, \tilde{X}) \leq \frac{1}{1-\alpha}\rho(K^{n_0}X, \tilde{K}^{n_0}X).$$

Evaluate the distance involved in the right-hand side of the last inequality, namely:

$$\begin{aligned}\rho(K^{n_0}X, \tilde{K}^{n_0}X) \leq\ & \rho(K(K^{n_0-1}X), \tilde{K}(K^{n_0-1}X)) \\ &+ \rho(\tilde{K}K(K^{n_0-2}X), \tilde{K}\tilde{K}(K^{n_0-2}X)) \\ &+ \rho(\tilde{K}^2K(K^{n_0-3}X), \tilde{K}^2\tilde{K}(K^{n_0-3}X)) \\ &+ \ldots + \rho(\tilde{K}^{n_0-1}(KX), \tilde{K}^{n_0-1}\tilde{K}X).\end{aligned}$$

The first term does not exceed δ; the second is no more than

$$4My_0\rho(K(K^{n_0-2}X), \tilde{K}(K^{n_0-2}X)) \le \frac{4My_0}{1!}\,\delta,$$

the third term is no greater than $(4My_0)^2\delta/2!$, and so on. As a result, we obtain

$$\rho(K^{n_0}X, \tilde{K}^{n_0}X) \le \delta\Big(1 + \frac{4My_0}{1!} + \frac{(4My_0)^2}{2!} + \ldots + \frac{(4My_0)^{n_0-1}}{(n_0-1)!}\Big) \le \delta e^{4My_0}.$$

Thus, small changes of the inverse problem data such that

$$\max_{t\in[0,2y_0]} \Big\{|h_1(t) - \tilde{h}_1(t)|, |h_2(t) - \tilde{h}_2(t)|\Big\} \le \delta$$

cause small changes of its solution:

$$|q(y) - \tilde{q}(y)| \le \rho(X, \tilde{X}) \le \frac{1}{1-\alpha}\, e^{4My_0}\delta.$$

1.3.4. Solution of the inverse problem by successive steps

Suppose that we have found the solution of the inverse problem for $y \in [0, y_0]$. Then the function $q(y)$ is known on this interval. Suppose that we know the functions $h_i(t)$ for $t \in [0, 2y_1]$, $y_1 > y_0$. Then, two equations (1.3.1), (1.3.2) form the linear Volterra system of integral equations relative to the functions $\Psi_1(y,t)$, $\Psi_2(y,t)$ in the trapezoid $\mathcal{D}_{y_0,y_1}$ (see Fig. 1.5).

It should be noted that below the characteristic $t+y = 2y_0$ the functions $\Psi_1(y,t)$ and $\Psi_2(y,t)$ are known; and equations (1.3.1), (1.3.2) hold automatically. Solving this system, we find, in particular, the values of $\vec{\Psi}(y_0,t)$ for $t \in [y_0, 2y_1 - y_0]$. Then we transfer the origin of the plane y,t to the point (y_0, y_0). As a result, we come to the inverse problem of the form (1.1.12)–(1.1.14), where $\Psi_1(y_0, t+y_0)$ and $\Psi_2(y_0, t+y_0)$ play the role of the functions h_1, h_2. Using the method discussed above, we can solve the inverse problem on an interval (y_0, y_0'), $y_0' \le y_1$. If we know *a priori* that the solution of the inverse problem exists, and $h_1(t)$, $h_2(t)$ are given for all t, then repeating this step, we can find the solution on each arbitrarily large interval $[0, y^*]$.

Note that in order to find $q(y)$ on the interval $[0, y^*]$, we use the values of $h_i(t)$ only on the interval $[0, 2y^*]$. This means that the methods proposed above are local.

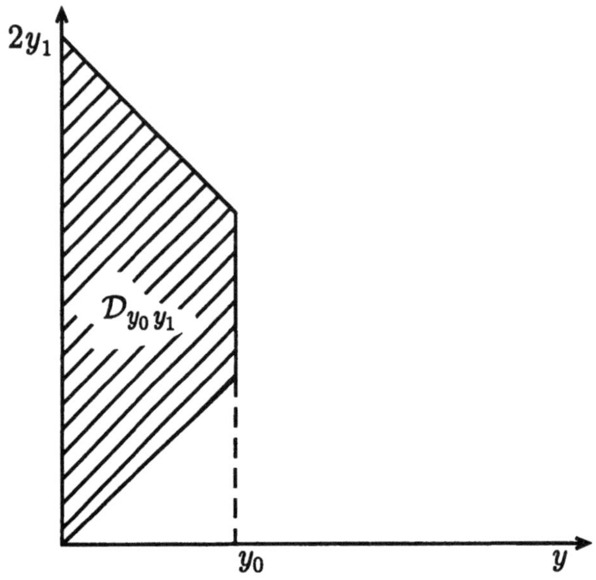

Fig. 1.5

1.3.5. The case of discontinuous $\sigma(y)$

The above considerations are applicable in the case when the function $q(y) = \sigma'/2\sigma$ is continuous or piecewise continuous. It occurs that after a modification we can apply them in the case when $\sigma(y)$ has a finite number of jump points. Suppose, for simplicity, that $\sigma(y)$ has a single jump point (if the number of jump points is more than one, we can apply this technique from one jump point to another).

We assume that $y_0 \geq 0$ is the point of jump. Let $\sigma^{\pm} = \sigma(y_0 \pm 0)$. The case of $y_0 = 0$ (meaningful in the scattering problem) was discussed above (see Subsection 1.2.3). We consider the case when $y_0 > 0$. Suppose, as before, that $\Psi_1(y,t)$ has singularity (1.2.1): $\Psi_1 \sim \delta(t-y)$ in a neighborhood of $t = y$, $y \in (0, y_0)$. Then, as pointed out in Subsection 1.2.3, from the properties of the inverse problem data we can recover: 1) the location of the point $y_0 > 0$, 2) the jump value (more precisely, σ^+/σ^-). Since the method is local, the jump has no influence on the procedure of determination of solution for $y < y_0$. Suppose that this solution is found, and the function $q(y)$ is known for $y < y_0$. Consider the domain $\mathcal{D}_{y_0 y_1}$ (see Fig. 1.5).

In this domain we may write out the linear system (1.3.1), (1.3.2) relative to the unknown functions $\Psi_1(y,t)$, $\Psi_2(y,t)$. However, in this case, the functions Ψ_1 and Ψ_2 have δ-singularities inside the domain $\mathcal{D}_{y_0 y_1}$ on the characteristic $t = 2y_0 - y$,

$$\Psi_2(y,t) = r\delta(t + y - 2y_0) + \Psi_2^*(y,t), \tag{1.3.7}$$

and on the characteristic $t = 2y_0 + y$,

$$\Psi_1(y,t) = r_1\delta(t - y - 2y_0) + \Psi_1^*(y,t). \tag{1.3.8}$$

Here Ψ_1^*, Ψ_2^* are classical (not generalized) functions, they may be discontinuous; r is the coefficient of reflection from the jump of $\sigma(y)$ at the point y_0. Following Subsection 1.2.3, the coefficient r is determined by the formula

$$r = -\frac{\sigma^+ - \sigma^-}{\sigma^+ + \sigma^-}.$$

The term with δ-singularity arises in (1.3.8) because the wave reflected from the point y_0 is reflected once again at the point 0; $r_1 = 0$ if the initial problem is the scattering problem and $y = 0$ is the continuity point of the function $\sigma(y)$. If in the scattering problem σ is discontinuous at zero, it is easy to show that

$$r_1 = r\,\frac{\sigma(+0) - \sigma(-0)}{\sigma(+0) + \sigma(-0)}.$$

Finally, if $y = 0$ is the string end and (1.1.4) holds, then $r_1 = r$.

In the domain $\mathcal{D}_{y_0 y_1}$ there is no other δ-singularity generated by the jump at the point y_0 if either $y_1 \le 2y_0$ or $r_1 = 0$. It is easy to make the solution have no δ-term with singularity on the characteristic $t = 2y_0 + y$. For this purpose, it suffices to change the boundary source (or the incident wave in the scattering problem) so that it will generate, parallel with the wave with δ-singularity (which corresponds to the first entry), the wave with singularity $2r_1\delta(t - y + y_0)$. This singularity extinguishes that in the wave reflected first at the point y_0 and then at zero. In this case it is not necessary to change the physical setting of the problem: the change can be done by formal mathematical manipulations. Indeed, we construct, instead of the function $\vec{\Psi}(y,t)$, the function

$$\vec{\Psi}^0(y,t) = \vec{\Psi}(y,t) - r_1\vec{\Psi}(y, t - 2y_0).$$

It is easy to see that $\vec{\Psi}^0(y,t)$ solves equation (1.1.12) and satisfies (1.1.13) and an analog of (1.1.14), namely:

$$\vec{\Psi}^0(y,t)|_{y=0} = \vec{H}(t) - r_1\vec{H}(t - 2y_0).$$

Evidently, the function $\vec{\Psi}^0$ is a solution with the required properties. To simplify the notations, we shall omit the superscript 0 of the function $\vec{\Psi}^0(y,t)$ in what follows.

The system of integral equations (1.3.1), (1.3.2), as before, holds; however, due to the δ-singularity of $\vec{\Psi}(y,t)$, there appears an additional term in equation (1.3.1). Let $t \neq 2y_0 - y$. Then $\vec{\Psi}^*(y,t) = \vec{\Psi}(y,t)$, and equation (1.3.1) becomes as follows:

$$\Psi_1^*(y,t) = \int_0^y q(\eta)\Psi_2^*(\eta,\eta+t-y)\,\mathrm{d}\eta + h_1(t-y) - re(y,t)q\Big(y_0 + \frac{y-t}{2}\Big).$$

The last term is connected with contribution of the δ-term to Ψ_2; $e(y,t)$ is equal to unity or zero depending on whether the characteristic along which we integrate intersects the δ-term of Ψ_2, namely: $e(y,t) = 1$ if $0 < y_0 + (y-t)/2 < y$, otherwise $e(y,t) = 0$. On changing from Ψ_i to Ψ_i^*, equation (1.3.2) remains invariable. The system so modified is a linear Volterra system of second kind. Solving this system, we find $\vec{\Psi}^*(y,t)$ for $y,t \in \mathcal{D}_{y_0y_1}$ and therefore $\vec{\Psi}(y_0-0,t)$ for $t \in (y_0, 2y_1 - y_0)$. Using (1.2.8), we find $\vec{\Psi}(y_0+0,t)$ for the same values of t. The values of $\vec{\Psi}(y_0+0,t)$, as before, can be considered as the inverse problem data for $y > y_0$.

The above technique can be evidently adapted to the case of any finite number of jump points of the function σ.

Note that at all the stages of solution of the inverse problem by the method suggested above, we have to solve either linear or nonlinear Volterra systems of integral equations of second kind. As is known, these equations can be solved by the method of successive approximations, which opens the way for the numerical solution of the problems.

Remark 1.3.2. As evidenced by the foregoing account, relation (1.2.4) holds, and hence[5]

$$q(y) = -2\Psi_2(y,y).$$

Using this equality, we can get rid of $q(y)$ in the setting of the problem by rewriting system (1.1.12) in the form

$$\vec{\Psi}_t + \Lambda\vec{\Psi}_y = -2\Psi_2(y,y)\begin{pmatrix} 0 & 1 \\ -1 & 0 \end{pmatrix}\vec{\Psi}. \tag{1.3.9}$$

[5]Hereafter we regard $\Psi_2(y,y)$ as $\Psi_2(y,y+0)$. The limit $\Psi_2(y,y-0)=0$ is evident. In general, if a function has a jump at a point and is identically equal to zero on one side from the point, then by the value of the function at the jump point is meant its limit value from the nontrivial side.

System (1.3.9) is a nonlinear hyperbolic system with a shifted argument. It holds for $t > y$. The inverse problem may be interpreted as the Cauchy problem for this system with initial data for $y = 0$.

Remark 1.3.3. The following difficulty arises when we pass to the multidimensional case. Suppose that we consider the inverse problem for the telegraph equation

$$u_{tt} - \Delta u = q(x)u, \qquad x = (x_1, x_2, x_3), \quad x_3 > 0,$$

with the conditions

$$u|_{t<0} = 0, \qquad u_{x_3}|_{x_3=0} = \delta(t)\delta(x_1, x_2), \qquad u|_{x_3=0} = f(x, t).$$

As is known (Romanov, 1987), $u|_{t<|x|} = 0$, and u is representable as

$$u = -\frac{1}{\pi}\delta(t^2 - |x|^2) + u_0(x, t),$$

where $u_0(x, t)$ is continuous for $t > |x|$:

$$u_0(x, |x| + 0) = -\frac{1}{4\pi|x|}\int_0^{|x|} q(\rho x/|x|)\,\mathrm{d}\rho.$$

Hence

$$q(x) = -\frac{4\pi}{|x|}(x, \nabla)(|x|u(x, |x| + 0)).$$

It is now clear that the problem in question can be reformulated as the problem for the nonlinear hyperbolic equation

$$u_{tt} - \Delta u + \Big(\frac{4\pi}{|x|}(x, \nabla)(|x|u(x, |x| + 0))\Big)u = 0$$

in the cone $t > |x|$ with the Cauchy data on the time-like surface

$$u_{x_3}|_{x_3=0} = 0, \qquad u|_{x_3=0} = f(x_1, x_2, t).$$

1.4. METHOD OF LINEAR INTEGRAL EQUATIONS

1.4.1. Fundamental system of solutions of equation (1.1.12)

We continue considering problem (1.1.12)–(1.1.14). Suppose that $\vec{\Theta}(y,t)$ is a solution of (1.1.12) for $y > 0$. We introduce a vector function

$$\vec{\Theta}^-(y,t) = \{\Theta_1^-(y,t), \Theta_2^-(y,t)\} = \{\Theta_2(y,-t), \Theta_1(y,-t)\}.$$

Let us show that $\vec{\Theta}^-(y,t)$ satisfies system (1.1.12) as well. Really,

$$\begin{aligned}
\Theta_{1t}^- + \Theta_{1y}^- &= \frac{\partial}{\partial t}\Theta_2(y,-t) + \frac{\partial}{\partial y}\Theta_2(y,-t) \\
&= -\dot{\Theta}_2(y,-t) + \frac{\partial}{\partial y}\Theta_2(y,-t) = q(y)\Theta_1(y,-t) = q(y)\Theta_2^-(y,t), \\
\Theta_{2t}^- - \Theta_{2y}^- &= \frac{\partial}{\partial t}\Theta_1(y,-t) - \frac{\partial}{\partial y}\Theta_1(y,-t) \\
&= -\dot{\Theta}_1(y,-t) - \frac{\partial}{\partial y}\Theta_1(y,-t) = -q(y)\Theta_2(y,-t) = -q(y)\Theta_1^-(y,t).
\end{aligned}$$

Here the point denotes the differentiation of the function Θ_i with respect to the second argument.

In what follows the vector function $\vec{\Theta}(y,t)$ will be thought of as a solution of system (1.1.12) for $y > 0$ subject to the condition

$$\vec{\Theta}(y,t)|_{y=0} = \begin{pmatrix} \delta(t) \\ 0 \end{pmatrix}. \tag{1.4.1}$$

Such a solution *a fortiori* exists and is unique. Note also that in system (1.1.12) the coordinates y and t are equivalent in the following sense. If we interchange the roles of the time and space coordinates, the system will continue describing the wave process with the speed of wave propagation equal to unity. Since the initial data (1.4.1) are concentrated at zero, therefore,

$$\vec{\Theta}(y,t)|_{|t|>y} = 0. \tag{1.4.2}$$

Besides (if $\sigma(y)$ has no discontinuities), we have

$$\vec{\Theta}(y,t) = \delta(t-y)\begin{pmatrix} 1 \\ 0 \end{pmatrix} + \begin{pmatrix} \theta_1(y,t) \\ \theta_2(y,t) \end{pmatrix} \tag{1.4.3}$$

where the functions $\theta_i(y,t)$ are piecewise continuous ($i=1,2$) and

$$\theta_2(y,y)=\frac{1}{2}q(y). \tag{1.4.4}$$

Evidently, the function

$$\vec{\Theta}^-(y,t)=\delta(t+y)\begin{pmatrix}0\\1\end{pmatrix}+\begin{pmatrix}\theta_2(y,-t)\\\theta_1(y,-t)\end{pmatrix} \tag{1.4.5}$$

solves system (1.1.12) and meets

$$\vec{\Theta}^-|_{y=0}=\begin{pmatrix}0\\\delta(t)\end{pmatrix}.$$

The functions $\vec{\Theta}$, $\vec{\Theta}^-$ form a fundamental system of solutions of (1.1.12) in the sense that, from $\vec{\Theta}$ and $\vec{\Theta}^-$, we can construct an explicit solution of (1.1.12) which corresponds to an arbitrary initial condition at $y=0$, in particular, $\vec{\Psi}(y,t)$. Really, introduce a convolution operation with respect to time as follows. Let $h=h(t)$ be a scalar function. The convolution $h*\vec{\Theta}$ is defined by the equality

$$h*\vec{\Theta}=\int_{-\infty}^{\infty}h(t-s)\vec{\Theta}(y,s)\,ds=\int_{-\infty}^{\infty}h(s)\vec{\Theta}(y,t-s)\,ds.$$

It is easy to verify that, for each $h(t)$, if $\vec{\Theta}$ satisfies system (1.1.12), so will $h*\vec{\Theta}$. If $\vec{\Theta}(y,t)$ solves the problem (1.1.12), (1.4.1), then it is clear that

$$(h*\vec{\Theta})(y,t)|_{y=0}=h*\begin{pmatrix}\delta(t)\\0\end{pmatrix}=\begin{pmatrix}h(t)\\0\end{pmatrix}.$$

Analogously, we have

$$(h*\vec{\Theta}^-)(y,t)|_{y=0}=\begin{pmatrix}0\\h(t)\end{pmatrix}.$$

1.4.2. Derivation of linear integral equations

Let $\vec{\Psi}(y,t)$ solve the problem (1.1.12)–(1.1.14), and let $H_1(t)$, $H_2(t)$ be the values of Ψ_1 and Ψ_2 for $y=0$. Then

$$\vec{\Psi}(y,t)=H_1*\vec{\Theta}(y,t)+H_2*\vec{\Theta}^-(y,t). \tag{1.4.6}$$

In the previous considerations of this section it did not matter whether $\sigma(y)$ was continuous or not. From here on, unless otherwise specified, we shall assume that $\sigma(y)$ is continuous. Then, by formula (1.2.7), we have

$$\vec{\Psi}(y,t) = \begin{pmatrix} \delta(t-y) \\ 0 \end{pmatrix} + \begin{pmatrix} \psi_1(y,t) \\ \psi_2(y,t) \end{pmatrix}$$

$$\vec{\Psi}(0,t) = \begin{pmatrix} H_1(t) \\ H_2(t) \end{pmatrix} = \begin{pmatrix} \delta(t) + h_1(t) \\ h_2(t) \end{pmatrix}. \tag{1.4.7}$$

Using equalities (1.4.6), (1.4.7), we obtain the following representation for $\vec{\Psi}(y,t)$:

$$\begin{aligned} \Psi_1(y,t) &= \int_{-\infty}^{\infty} (\delta(t-s) + h_1(t-s))(\delta(s-y) + \theta_1(y,s))\,\mathrm{d}s \\ &\quad + \int_{-\infty}^{\infty} h_2(t-s)\theta_2(y,-s)\,\mathrm{d}s \\ &= \delta(t-y) + h_1(t-y) + \theta_1(y,t) \\ &\quad + \int_{-\infty}^{\infty} h_1(t-s)\theta_1(y,s)\,\mathrm{d}s + \int_{-\infty}^{\infty} h_2(t+s)\theta_2(y,s)\,\mathrm{d}s \\ \Psi_2(y,t) &= \int_{-\infty}^{\infty} (\delta(t-s) + h_1(t-s))\theta_2(y,s)\,\mathrm{d}s \\ &\quad + \int_{-\infty}^{\infty} h_2(t-s)(\delta(s+y) + \theta_1(y,-s))\,\mathrm{d}s \\ &= h_2(t+y) + \theta_2(y,t) \\ &\quad + \int_{-\infty}^{\infty} h_1(t-s)\theta_2(y,s)\,\mathrm{d}s + \int_{-\infty}^{\infty} h_2(t+s)\theta_1(y,s)\,\mathrm{d}s. \end{aligned} \tag{1.4.8}$$

Formally, in both the equalities we integrate in infinite limits. However, in reality, 1) $h_i \equiv 0$ for $t < 0$; 2) $\theta_i(y,t) \equiv 0$ for $|t| > y$. Taking this into account, we can rewrite formula (1.4.8). Suppose that $t > y$. Then

$$\begin{aligned} \Psi_1(y,t) &= h_1(t-y) + \int_{-y}^{-t} h_1(t-s)\theta_1(y,s)\,\mathrm{d}s + \int_{-t}^{y} h_2(t-s)\theta_2(y,s)\,\mathrm{d}s, \\ \Psi_2(y,t) &= h_2(t+y) + \int_{-y}^{t} h_1(t-s)\theta_2(y,s)\,\mathrm{d}s + \int_{-y}^{y} h_2(t+s)\theta_1(y,s)\,\mathrm{d}s. \end{aligned} \tag{1.4.9}$$

The relation for $t \in (-y, y)$ which follows from (1.4.8) is more important for us. Note that, since the wave propagation has a finite speed (in the coordinates y, t it is equal to unity), therefore,

$$\Psi_1(y,t) \equiv \Psi_2(y,t) \equiv 0, \quad \text{for} \quad -y < t < y,$$

whence

$$\theta_1(y,t) + \int_{-y}^{t} h_1(t-s)\theta_1(y,s)\,\mathrm{d}s + \int_{-t}^{y} h_2(t+s)\theta_2(y,s)\,\mathrm{d}s = 0,$$
$$\theta_2(y,t) + \int_{-y}^{t} h_1(t-s)\theta_2(y,s)\,\mathrm{d}s + \int_{-t}^{y} h_2(t+s)\theta_1(y,s)\,\mathrm{d}s + h_2(t+y) = 0. \tag{1.4.10}$$

Formulas (1.4.10) represent a system of linear integral equations of second kind relative to the functions $\theta_1(y,t)$, $\theta_2(y,t)$. More precisely, the vector function $\vec{\theta}(y,t) = \{\theta_1(y,t), \theta_2(y,t)\}$ satisfies this system as a function of t for each fixed y. The variable y plays the role of a parameter.

System (1.4.10) and related equations can be conveniently written in a concise form. To this end, we introduce operators $\hat{H}_1$ and $\hat{H}_2$ as

$$\hat{H}_1(\varphi(t)) = \int_{-y}^{t} h_1(t-s)\varphi(s)\,\mathrm{d}s, \qquad \hat{H}_2(\varphi(t)) = \int_{-t}^{y} h_2(t+s)\varphi(s)\,\mathrm{d}s.$$

It should be noted that the operators $\hat{H}_1$ and $\hat{H}_2$ depend on the parameter y (though y is not involved). So, system (1.4.10) can be rewritten as follows

$$\theta_1 + \hat{H}_1\theta_1 + \hat{H}_2\theta_2 = 0,$$
$$\theta_2 + \hat{H}_1\theta_2 + \hat{H}_2\theta_1 + h_2(y+t) = 0.$$

We see that the operator $\hat{H}_1$ is a Volterra integral operator, and $\hat{H}_2$ is a Fredholm operator with a symmetric kernel. The last fact becomes evident if we take into account that

$$h_2|_{t<0} \equiv 0$$

and write the operator $\hat{H}_2$ in the form

$$\hat{H}_2\varphi(t) = \int_{-y}^{y} h_2(t+s)\varphi(s)\,\mathrm{d}s.$$

1.4.3. Recovery of the coefficient $q(y)$ from the solution of the linear integral equation

If system (1.4.10) is solved and the function $\vec{\theta}(y,t)$ is found, we can recover $q(y)$ easily. We shall give two ways of this recovery.

1. Using formula (1.4.4), we obtain

$$q(y) = 2\theta_2(y,y).$$

2. Calculate

$$\vec{\Theta}^0(y) \stackrel{\text{def}}{=} \int_{-\infty}^{\infty} \vec{\Theta}(y,t)\,\mathrm{d}t = \begin{pmatrix} \displaystyle\int_{-\infty}^{\infty} \delta(t-y)\,\mathrm{d}t + \int_{-\infty}^{\infty} \theta_1(y,t)\,\mathrm{d}t \\ \displaystyle\int_{-\infty}^{\infty} \theta_2(y,t)\,\mathrm{d}t \end{pmatrix}$$

$$= \begin{pmatrix} 1 + \displaystyle\int_{-y}^{y} \theta_1(y,t)\,\mathrm{d}t \\ \displaystyle\int_{-y}^{y} \theta_2(y,t)\,\mathrm{d}t \end{pmatrix} = \begin{pmatrix} \Theta_1^0(y) \\ \Theta_2^0(y) \end{pmatrix}.$$

We can obtain the system of differential equations for the function $\vec{\Theta}^0(y)$:

$$\frac{\partial \Theta_1^0}{\partial y} = q(y)\Theta_2^0, \qquad \frac{\partial \Theta_2^0}{\partial y} = q(y)\Theta_1^0. \tag{1.4.11}$$

Really, differentiating Θ_1^0 and using system (1.1.12), we obtain

$$\frac{\partial}{\partial y}\Theta_1^0(y) = \int_{-\infty}^{\infty} \frac{\partial}{\partial y}\Theta_1(y,t)\,\mathrm{d}t$$

$$= -\int_{-\infty}^{\infty} \frac{\partial}{\partial t}\Theta_1(y,t)\,\mathrm{d}t + q(y)\int_{-\infty}^{\infty} \Theta_2(y,t)\,\mathrm{d}t = q(y)\Theta_2^0,$$

because

$$\int_{-\infty}^{\infty} \frac{\partial}{\partial t}\Theta_1(y,t)\,\mathrm{d}t = 0$$

due to the finiteness of $\theta_1(y,t)$. The second relation (1.4.11) is obtained analogously. Introducing the functions

$$\chi_1^0(y) = \Theta_1^0(y) + \Theta_2^0(y), \qquad \chi_2^0(y) = \Theta_1^0(y) - \Theta_2^0(y),$$

we derive from (1.4.11) the following equations for the functions $\chi_i^0(y)$:

$$\frac{\partial \chi_1^0}{\partial y} = q(y)\chi_1^0(y), \qquad \frac{\partial \chi_2^0}{\partial y} = -q(y)\chi_2^0(y). \tag{1.4.12}$$

Relations (1.4.12) are useful to obtain an expression for $\sigma(y)$. Indeed, taking into account $q(y) = \sigma'(y)/(2\sigma(y))$ and integrating the equalities

$$\frac{\sigma'(y)}{\sigma(y)} = 2\,\frac{\chi_1^{0\prime}(y)}{\chi_1^0(y)} = -2\,\frac{\chi_2^{0\prime}(y)}{\chi_2^0(y)},$$

we arrive at

$$\sigma(y) = \sigma(0)(\chi_1^0(y))^2 = \sigma(0)(\chi_2^0(y))^{-2}. \tag{1.4.13}$$

Here we have taken into account that

$$\chi_1^0(0) = \chi_2^0(0) = 1$$

owing to the initial conditions for $\vec{\Theta}$.

1.4.4. Structure of equations (1.4.10). Existence and uniqueness of solution

Consider system (1.4.10). Evidently, after a simple transformation it is decomposed into two independent equations. Really, let

$$\chi_1(y,t) = \theta_1(y,t) + \theta_2(y,t), \qquad \chi_2(y,t) = \theta_1(y,t) - \theta_2(y,t).$$

Summing and subtracting equations of system (1.4.10), we find

$$\chi_1 + \hat{H}_1\chi_1 + \hat{H}_2\chi_1 + h_2(t+y) = 0, \tag{1.4.14}$$

$$\chi_2 + \hat{H}_1\chi_2 - \hat{H}_2\chi_2 - h_2(t+y) = 0, \tag{1.4.15}$$

or, in the expanded form,

$$\chi_1(y,t) + \int_{-y}^{t} h_1(t-s)\chi_1(y,s)\,\mathrm{d}s + \int_{-t}^{y} h_2(t+s)\chi_1(y,s)\,\mathrm{d}s + h_2(t+y) = 0, \tag{1.4.16}$$

$$\chi_2(y,t) + \int_{-y}^{t} h_1(t-s)\chi_2(y,s)\,\mathrm{d}s - \int_{-t}^{y} h_2(t+s)\chi_2(y,s)\,\mathrm{d}s - h_2(t+y) = 0. \tag{1.4.17}$$

Each of equations (1.4.14)–(1.4.17) has only one unknown function ($\chi_1(y,t)$ or $\chi_2(y,t)$).

Evidently,

$$\chi_i^0(y) = 1 + \int_{-y}^{y} \chi_i(y,t)\,\mathrm{d}t, \qquad i = 1, 2,$$

where the functions χ_1^0, χ_2^0 are those introduced above. Therefore, in order to solve the inverse problem, it suffices to solve any one of equations (1.4.14)–(1.4.17). Then we can calculate the corresponding function χ_i^0 and find $\sigma(y) = \sigma(0)(\chi_i^0)^{\pm 2}$ (the exponent is equal to 2 for $i = 1$ and to -2 for $i = 2$).

Naturally, the following questions arise: 1) Which restrictions should be imposed on the inverse problem data $h_1(t)$, $h_2(t)$ to guarantee existence of a solution? 2) Is the solution unique?

It occurs that the answer is indivisible. The necessary and sufficient condition for a solution of the inverse problem to exist on the interval $(0, y_0)$ is that the homogeneous equations which correspond to equations (1.4.14), (1.4.15) have only the trivial solution for all $y \in (0, y_0)$. This condition evidently provides also uniqueness of the solution.

So, the necessary and sufficient condition for the unique solvability of the inverse problem is that for all y $(0 < y < y_0)$ there are no nontrivial solutions of the equations

$$\varphi(t) + \int_{-y}^{t} h_1(t-s)\varphi(s)\,ds + \int_{-t}^{y} h_2(t+s)\varphi(s)\,ds = 0, \qquad (1.4.18)$$

$$\varphi(t) + \int_{-y}^{t} h_1(t-s)\varphi(s)\,ds - \int_{-t}^{y} h_2(t+s)\varphi(s)\,ds = 0. \qquad (1.4.19)$$

In other words, for none of $y \in (0, y_0)$, zero is an eigenvalue of the operators.

This assertion will be proved in Subsections 1.4.5–1.4.7. Recall that (1.4.18), (1.4.19) are two independent equations. Note also that the inverse problems with different data, i. e., different pairs of functions $H_1(t)$, $H_2(t)$, may have the same solution. Really, the function $H_1(t)$ simulates the wave issuing from the point $y = 0$ to the right: $H_1(t) = \Psi_1(y,t)|_{y=0}$. This wave, as mentioned above, either is generated by a source located at the point $y = 0$ or is coming from the domain $y < 0$. It is clear that in both the cases, for a fixed equation (1.1.12), this wave may be arbitrarily changed. This obviously causes some changes in the wave $\Psi_2(0,t) = H_2^-(t)$ reflected from the medium nonhomogeneities. Consider the particular case when $H_1(t) = \delta(t)$. Denote the corresponding solution $\vec{\Psi}(y,t)$ of the direct problem by $\vec{\Psi}^0(y,t)$, $\Psi_2^0(y,t)|_{y=0} = h_2^0(t)$, and the operator $\hat{H}_2$ generated by the function h_2^0 by $\hat{H}_2^0$. In this case equations (1.4.14), (1.4.15) will take the form

$$\chi_1 + \hat{H}_2^0\chi_1 + h_2^0(t+y) = 0, \qquad (1.4.20)$$

$$\chi_2 - \hat{H}_2^0\chi_2 - h_2^0(t+y) = 0. \qquad (1.4.21)$$

Equations (1.4.20), (1.4.21) are integral equations with a symmetric kernel. Hence, all the eigenvalues of the operator $\hat{H}_2^0$ are real. The conditions for the inverse problem (1.1.12)–(1.1.14) to have a unique solution on the interval $[0, y_0]$ mean, in this situation, that for none of $y \in [0, y_0]$ the numbers $+1$, -1 are eigenvalues of the operator $\hat{H}_2^0$.

Therefore to solve equations (1.4.20), (1.4.21), we can apply the method of successive approximations. Really, for y sufficiently small, the operator $\hat{H}_2^0$ is small in norm, because it is an integral operator and the integration is over a small interval. All eigenvalues of the operator $\hat{H}_2^0$, for y sufficiently small, necessarily lie on the interval $(-1, 1)$. It is easy to verify that all the eigenvalues of the operator $\hat{H}_2^0$ depend continuously on y; therefore, if the conditions of the solution existence for $y \in [0, y_0]$ hold, they remain on the interval $(-1, 1)$, even though y_0 is arbitrarily large. Therefore, the method of successive approximations is applicable.

1.4.5. Modification of equations (1.4.14), (1.4.15) under a change of the source

Let us clarify the connection between equations (1.4.18), (1.4.19) and (1.4.14), (1.4.15). Suppose that $\vec{\Psi}^0(y,t)$ is the above-mentioned solution of the direct problem for system (1.1.12) which meets the trivial initial condition

$$\vec{\Psi}^0(y,t)|_{t<0} = 0 \tag{1.4.22}$$

and the boundary condition

$$\Psi_1^0|_{y=0} = \delta(t). \tag{1.4.23}$$

Let a function $\vec{\Psi}(y,t)$ be a solution of (1.1.12) satisfying the initial condition (1.4.22) and the boundary condition

$$\Psi_1(y,t)|_{y=0} = H_1(t) \tag{1.4.24}$$

with an arbitrary function $H_1(t)$ vanishing for $t < 0$. The solution of the problem (1.1.12), (1.4.22), (1.4.24) is expressed in terms of $H_1(t)$ and $\vec{\Psi}^0(y,t)$ by the convolution

$$\vec{\Psi}(y,t) = \int_{-\infty}^{\infty} H_1(t-s)\vec{\Psi}^0(y,s)\,ds = H_1 * \vec{\Psi}^0(y,t).$$

In particular, denoting $\Psi_2(0,t) = H_2(t)$ and taking into account that $h_2^0(t) = \Psi_2^0(0,t)$, we obtain

$$H_2(t) = H_1(t) * h_2^0(t). \tag{1.4.25}$$

In the inverse problem, the functions $H_1(t)$ and $H_2(t)$ are known. Therefore, relation (1.4.25) is an equation relative to $h_2^0(t)$. If, as earlier,

$$H_1(t) = \delta(t) + h_1(t), \qquad H_2(t) = h_2(t),$$

then equation (1.4.25) is a Volterra integral equation of second kind with the difference kernel

$$h_2(t) = \int_0^t h_1(t-s)h_2^0(s)\,\mathrm{d}s + h_2^0(t). \tag{1.4.26}$$

It is not so hard to solve this equation. If $H_1(t)$ is a continuous or piecewise continuous function, then equation (1.4.23) is of first kind. The problem of determination of its solution is ill-posed. We shall not present here methods for solving equations of this sort. For a detailed discussion of this subject see, for example, Tikhonov and Arsenin (1977). Using equality (1.4.26), we can easily obtain a relation between the three operators $\hat{H}_1$, $\hat{H}_2$, and $\hat{H}_2^0$, namely:

$$\hat{H}_2 = (E + \hat{H}_1)\hat{H}_2^0$$

(E is the unit operator). It is readily seen that

$$h_2(t+y) = (E + \hat{H}_1)(h_2^0(t+y)).$$

The two last formulas mean that equations (1.4.14), (1.4.15) can be rewritten as

$$(E + \hat{H}_1)(\chi_i \pm \hat{H}_2^0\chi_i \pm h_2^0(y+t)) = 0 \tag{1.4.27}$$

($i = 1, 2$; plus corresponds to $i = 1$; minus to $i = 2$). Since the operator $\hat{H}_1$ is a Volterra integral operator, the operator $E + \hat{H}_1$ is invertible. Therefore, equations (1.4.14), (1.4.15) are equivalent to (1.4.20), (1.4.21). Thus, to prove the existence theorem to a depth of y_0 it suffices to show that the criterion of the solvability of the inverse problem to a depth of y_0 is the invertibility of the operators $E \pm \hat{H}_2^0$ for each $y \in (0, y_0)$.

1.4.6. Proof of necessity. Conditions for solvability of the inverse problem

Assume that the inverse problem for equation (1.1.12) subject to conditions (1.4.22) and

$$\vec{\Psi}^0|_{y=0} = \begin{pmatrix} \delta(t) \\ h_2^0(t) \end{pmatrix} \tag{1.4.28}$$

is solvable ($y \in (0, y_0)$). Let the condition of invertibility of one of the operators $E \pm \hat{H}_2^0$ be violated for a certain $y^* \in (0, y_0)$; i.e., there exists a function $p(t)$ ($t \in (-y^*, y^*)$) satisfying one of the equations

$$p \pm \hat{H}_2^0 p = 0.$$

We assume that $p(t)$ is equal to zero for $t \notin (-y^*, y^*)$. Suppose, for definiteness, that $p - \hat{H}_2^0 p = 0$. We construct the function $\vec{\Phi}(y,t)$ as follows:

$$\vec{\Phi}(y,t) = \vec{\Psi}^0(y,t) * p^-(t) - \vec{\Psi}^{0-}(y,t) * p(t). \tag{1.4.29}$$

Here

$$\vec{\Psi}^{0-}(y,t) = \begin{pmatrix} \Psi_2^0(y,-t) \\ \Psi_1^0(y,-t) \end{pmatrix}, \qquad p^-(t) := p(-t).$$

Evidently, $\vec{\Phi}(y,t)$ solves system (1.1.12). Suppose that $y = 0$, $t \in (-y^*, y^*)$. Then

$$\vec{\Phi}(y,t)|_{y=0} = \begin{pmatrix} p(-t) - \int_{-y}^{y} h_2^0(\tau - t) p(\tau)\,\mathrm{d}\tau \\ \int_{-y}^{y} h_2^0(t-\tau) p(-\tau)\,\mathrm{d}\tau - p(t) \end{pmatrix}$$

$$= \begin{pmatrix} p(t') - \int_{-y}^{y} h_2^0(\tau + t') p(\tau)\,\mathrm{d}\tau \Big|_{t'=-t} \\ \int_{-y}^{y} h_2^0(t+\tau) p(\tau)\,\mathrm{d}\tau - p(t) \end{pmatrix} = 0.$$

Since system (1.1.12) is hyperbolic, we can guarantee that

$$\vec{\Phi}(y,t) \equiv 0 \quad \text{for} \quad y - y^* \le t \le y^* - y \quad (y < y^*).$$

On the other hand, taking into account (1.4.29), we obtain

$$\Phi_1(y,t) = p(y-t) + \int_{y-t}^{y^*} \psi_1^0(y, t+\tau) p(\tau)\,\mathrm{d}\tau - \int_{y+t}^{y^*} \psi_2^0(y, \tau - t) p(\tau)\,\mathrm{d}\tau = 0$$

for $t \in (y - y^*, y^* - y)$. Setting $y - t = s$ and $y + t = y^*$ in the last equality, we get

$$p(s) = -\int_s^{y^*} \Psi_1^0\Big(\frac{y^*+s}{2}, \frac{y^*-s}{2} + \tau\Big) p(\tau)\,\mathrm{d}\tau, \qquad s \in (-y^*, y^*).$$

This is a homogeneous linear Volterra equation of second kind relative to $p(s)$. This equation yields $p(s) \equiv 0$. Analogously, we conclude that if p solves the equation $p + \hat{H}_2^0 p = 0$, then $p \equiv 0$. The necessity is shown.

1.4.7. Proof of sufficiency

Suppose that for each $y \in (0, y_0)$, zero is not an eigenvalue of the operators $(E \pm \hat{H}_2^0)$. Hence, equations (1.4.20), (1.4.21) are solvable for each y. Their solutions are $\chi_1(y,t)$, $\chi_2(y,t)$. Note that the functions χ_1 and χ_2 are differentiable with respect to y and t for $|t| < y$ if $h_2^0(t)$ is sufficiently smooth. It is easy to verify that

$$\frac{\partial}{\partial t}\hat{H}_2^0\chi_i(y,t) = -\hat{H}_2^0\dot{\chi}_i(y,t) + h_2^0(t+y)\chi_i(y,y),$$

$$\frac{\partial}{\partial y}\hat{H}_2^0\chi_i(y,t) = \hat{H}_2^0\chi_i'(y,t) + h_2^0(t+y)\chi_i(y,y)$$

where the prime and the dot denote the derivatives with respect to y and t, respectively. Differentiating equation (1.4.20) with respect to t, equation (1.4.21) with respect to y and summing the resulting equations, we obtain

$$\dot{\chi}_1 + \chi_2' - \hat{H}_2^0(\dot{\chi}_1 + \chi_2') + h_2^0(t+y)(\chi_1(y,y) - \chi_2(y,y)) = 0.$$

We see that $\dot{\chi}_1 + \chi_2'$ as a function of t satisfies the integral equation essentially similar to (1.4.21), with the only difference that the last term has the multiplier $\chi_1(y,y) - \chi_2(y,y)$ constant for a fixed y. Since equation (1.4.21) is uniquely solvable, we obtain

$$\dot{\chi}_1 + \chi_2' = -q\chi_2$$

where

$$q(y) := \chi_1(y,y) - \chi_2(y,y).$$

Analogously we obtain the second equation

$$\dot{\chi}_2 + \chi_1' = q\chi_1.$$

Hence, we infer that

$$\vec{\theta} = (\theta_1, \theta_2) \qquad \left(\theta_1 = \frac{1}{2}(\chi_1 + \chi_2), \quad \theta_2 = \frac{1}{2}(\chi_1 - \chi_2)\right)$$

satisfies system (1.1.12) for $|t| < y$, with $q(y) = 2\theta_2(y,y)$. We construct now $\vec{\Theta}(y,t)$ extending the functions $\theta_1(y,t)$, $\theta_2(y,t)$ trivially for $|t| > y$ $(0 < y < y_0)$ and setting

$$\vec{\Theta}(y,t) = \begin{pmatrix} \delta(t-y) + \theta_1(y,t) \\ \theta_2(y,t) \end{pmatrix}.$$

It is easy to verify that $\vec{\Theta}$ is a solution of the Cauchy problem for system (1.1.12) with condition (1.4.1). Finally, the function $\vec{\Psi}(y,t)$ is determined from formula (1.4.6). The function $\vec{\Psi}(y,t)$ will satisfy system (1.1.12) and conditions (1.4.1) by the construction. From the system (1.4.20), (1.4.21) it follows that $\vec{\Psi}(y,t)$ satisfies also the initial condition (1.1.13). Sufficiency is proved.

1.5. THE CASE OF DISCONTINUOUS $\sigma(y)$

1.5.1. The first method

For definiteness, we consider the case when $h_1(t) \equiv 0$, $h_2(t) = h_2^0(t)$. Note that equations (1.4.18), (1.4.19) hold in the case of discontinuous $\sigma(y)$, however, their properties are changed.

Suppose, for simplicity, that we have only one jump point of $\sigma(y)$: $y = y_0 > 0$. Then, to obtain a solution for $y < y_0$, we can use the results of Section 1.4 (the methods are local). As indicated above, the function $h_2^0(t)$ is continuous or piecewise continuous for $0 < t < 2y_0$. At the point $t = 2y_0$, the function $h_2^0(t)$ has the δ-singularity:

$$h_2^0(t) = r\delta(t - 2y_0) + \tilde{h}_2^0(t)$$

where $\tilde{h}_2^0$ is a piecewise continuous function.

Suppose that $y > y_0$. Then equations (1.4.18), (1.4.19) are as follows:

$$\chi_i(y,t) \pm \int_{-t}^{y} \tilde{h}_2^0(t+s)\chi_i(y,s)\,\mathrm{d}s \pm r\chi_i(y, 2y_0 - t)\varepsilon(t + y - 2y_0)$$
$$\pm \tilde{h}_2^0(y+t) \pm r\delta(2y_0 - t - y) = 0.$$

Hereafter, in deciding between "+" or "−", we choose the upper sign for equation (1.4.18) and the lower sign for equation (1.4.19). The solution of the last equation should be sought in the form

$$\chi_i(y,t) = \mp r\delta(2y_0 - t - y) + \tilde{\chi}_i(y,t).$$

Then, for the function $\tilde{\chi}_i$, we obtain the equation

$$\tilde{\chi}_i(y,t) \pm \int_{-t}^{y} \tilde{h}_2^0(t+s)\tilde{\chi}_i(y,s)\,\mathrm{d}s - r\tilde{h}_2^0(t + 2y_0 - y)$$
$$\pm r\varepsilon(t + y - 2y_0)\tilde{\chi}_i(y, 2y_0 - t) \pm h_i^0(y+t) = 0.$$

This is a linear integro-functional equation relative to the function $\tilde{\chi}_i(y,t)$. It may be easily reduced to a linear Fredholm equation of second kind. We shall not investigate this equation more closely because the second method seems to be more preferable.

1.5.2. The second method

This method is essentially similar to that presented in Subsection 1.3.5:

1) Using the inverse problem solved for $y < y_0$, we construct the solution $\vec{\Psi}(y,t)$ of the direct problem for $y \in (0, y_0)$.

2) We find the value of $\vec{\Psi}(y+0,t)$ by formula (1.2.8).

3) Transferring the origin to the point (y_0, y_0) and interpreting the found values

$$\vec{\Psi}(y_0+0, y_0+t) = \vec{H}^{(1)}(t)$$

as data of a new inverse problem of the type (1.1.12)–(1.1.14), we solve this inverse problem using any method discussed above.

The advantageous distinction of this method from that considered in Subsection 1.3.5 is that if both equations (1.4.14), (1.4.15) (or (1.4.18), (1.4.19)) are solved, then from the obtained functions $\chi_1(y,t)$, $\chi_2(y,t)$ or, which is equivalent, from $\vec{\theta}(y,t)$ we recover the solution $\vec{\Psi}(y,t)$ for $y \in (0, y_0)$ by means of quadratures. To this end, it suffices to apply formulas (1.4.9). In this case, we need not solve any additional integral equations.

1.6. THE GEL'FAND – LEVITAN EQUATION FOR SECOND-ORDER HYPERBOLIC EQUATIONS

1.6.1. Derivation of a linear integral equation

As noted above, reduction of the string equation (1.1.1) to system (1.1.12) of first-order equations is not so principal. It is possible (in various ways) to reduce the inverse problem both to a nonlinear Volterra system and to a linear integral equation in terms connected with a second-order equation. We shall demonstrate now one of the variants of reduction of an equation slightly more general than (1.1.1) or the equivalent equation (1.1.8).

Suppose that the function $U(y,t)$ is a solution of the problem

$$U_{tt} = \frac{1}{\sigma}\frac{\partial}{\partial y}(\sigma U_y) + qU, \qquad y > 0, \tag{1.6.1}$$

$$U|_{t<0} = 0, \tag{1.6.2}$$

$$\sigma \frac{\partial U}{\partial y}\bigg|_{y=0} = \delta(t) \tag{1.6.3}$$

where $\sigma(y)$ and $q(y)$ are continuous functions of y; $\sigma(y) > 0$ is a piecewise smooth function. We suppose that

$$U|_{y=0} = f(t)$$

where $f(t)$ is a given function. We need to determine one of the functions $\sigma(y)$, $q(y)$ if the other is known.

Introduce a function $U^+(y,t)$ as the odd extension of the function $U(y,t)$ with respect to t:

$$U^+(y,t) \equiv U(y,t) \quad \text{for} \quad t > 0, \qquad U^+(y,-t) = -U^+(y,t).$$

The function U^+ satisfies equation (1.6.1) and the boundary conditions

$$\sigma \left.\frac{\partial U^+}{\partial y}\right|_{y=0} = \delta(t) - \delta(t) = 0,$$

$$U^+|_{y=0} = f^+(t)$$

where $f^+(t)$ is the odd extension of the function $f(t)$.

We introduce, parallel with $U(y,t)$, a function $V(y,t)$ which solves equation (1.6.1) subject to the conditions

$$V(y,t)|_{y=0} = \delta(t),$$

$$\sigma \left.\frac{\partial V}{\partial y}\right|_{y=0} = 0. \tag{1.6.4}$$

Such a solution $V(y,t)$ exists, is unique and equal to zero identically[6] for $t > y$; it is even in t and has δ-singularities for $t = \pm y$:

$$V(y,t) = \frac{1}{2}\sqrt{\frac{\sigma(0)}{\sigma(y)}}(\delta(t-y) + \delta(t+y) + v(y,t)). \tag{1.6.5}$$

Here the function $v(y,t)$ is continuous for $-y < t < y$ and even in t; $v \equiv 0$ for $|t| > y$. If the function V is known, then by this function, using the convolution, we can construct the function $U^+(y,t)$:

$$U^+(y,t) = f^+(t) * V(y,t) = \int_{-\infty}^{\infty} V(y,s) f^+(t-s)\,\mathrm{d}s. \tag{1.6.6}$$

Taking into account the properties of $V(y,t)$ and the fact that $U^+(y,t) \equiv 0$ for $|t| < y$, we obtain from (1.6.6) that, for $|t| < y$,

$$0 = \int_{-\infty}^{\infty} V(y,s) f(t-s)\,\mathrm{d}s = \sqrt{\frac{\sigma(0)}{\sigma(y)}}\Big(f(t-y) + f(t+y) + \int_{-y}^{y} v(y,s) f(t-s)\,\mathrm{d}s\Big).$$

[6]Evidently, the function V may be interpreted as a solution of equation (1.6.1) with initial data for $y = 0$. In this case, the coordinates y and t interchange their roles: y plays the role of time; and t, of spatial coordinate. The initial data for V, when $y = 0$, are concentrated at the point $t = 0$. Since the speed of the disturbance propagation remains unit after interchanging the roles of y and t, this implies that $V|_{|t|>y} \equiv 0$.

Hence,

$$\int_{-y}^{y} v(y,s)f^+(t-s)\,\mathrm{d}s + f^+(t-y) + f^+(t+y) = 0.$$

Since v is even and f^+ is odd, the last equality may be rewritten for $t \in (0, y)$ in the form

$$\int_0^y v(y,s)(f^+(t-s) + f(t+s))\,\mathrm{d}s + f(t+y) - f(t-y) = 0. \qquad (1.6.7)$$

Equation (1.6.7) is a linear integral equation of the first kind. It may be reduced to an equation of the second kind. Really, the function $U(y,t)$ has a singularity of the jump type for $y = t$. This jump is determined by the formula

$$U(y,t) = -\frac{1}{\sqrt{\sigma(0)\sigma(y)}}\,\varepsilon(t-y) + \ldots$$

where the dots stand for a term continuous for $y = t$. Substituting $y = 0$ in this formula, we obtain

$$f(t) = -\frac{1}{\sigma(0)}\,\varepsilon(t) + \ldots. \qquad (1.6.8)$$

Thus, we find $\sigma(0)$ from $f(0)$. The jump of the function $f(t)$ at zero is equal to $-\sigma^{-1}(0)$; while the jump of $f^+(t)$ is

$$[f^+(t)]|_{t=0} = -\frac{2}{\sigma(0)}.$$

Hence, the derivative of $f^+(t)$ with respect to t is

$$\frac{\mathrm{d}f(t)}{\mathrm{d}t} = -\frac{2}{\sigma(0)}\,\delta(t) + \dot{f}(|t|) \qquad (1.6.9)$$

where $\dot{f}(t)$ is a piecewise continuous function, which coincides with $\mathrm{d}f/\mathrm{d}t$ on the intervals of $f(t)$ continuity.

Differentiating equation (1.6.7) with respect to t and taking into account the δ-singularity of $\mathrm{d}f^+/\mathrm{d}t$, we obtain

$$-\frac{2}{\sigma(0)}\,v(y,t) + \int_0^y (\dot{f}(|t-s|) + \dot{f}(t+s))v(y,s)\,\mathrm{d}s + \dot{f}(y+t) + \dot{f}(y-t). \quad (1.6.10)$$

Equation (1.6.10) is a linear integral Fredholm equation of second kind with a symmetric kernel. More precisely, we have obtained the set of such equations depending on the parameter y. From the theoretical point of view, (1.6.10) plays a similar role as equations (1.4.14), (1.4.15) for problem (1.1.12)–(1.1.14), namely:

a) the solvability criterion of the inverse problem for equation (1.6.1) is formulated through (1.6.10) as follows: none of y gives the spectrum in equation (1.6.10);

b) this criterion provides also the uniqueness of the solution.

1.6.2. Recovery of coefficients by solution of the Gel'fand – Levitan equation

As above, investigating the singularity of the solution V, we find that

$$v|_{t=y} = \frac{1}{2}\frac{\sigma'(0)}{\sigma(0)} - \frac{1}{2}\int_0^y \Big(q(\eta) - \frac{(\sqrt{\sigma(\eta)})''}{\sqrt{\sigma(\eta)}}\Big)\, \mathrm{d}\eta, \tag{1.6.11}$$

whence

$$2\frac{\mathrm{d}}{\mathrm{d}y}v(y,y) = \frac{(\sqrt{\sigma(y)})''}{\sqrt{\sigma(y)}} - q(y). \tag{1.6.12}$$

If σ is given, then by formula (1.6.12) we derive $q(y)$. If $q(y)$ is given, then (1.6.12) is a linear differential equation of the second order relative to $\sqrt{\sigma(y)}$. Let us find the initial conditions for this equation. From (1.6.8) it follows that

$$\sqrt{\sigma}|_{y=0} = (-f(0))^{-1/2},$$

and (1.6.11) yields

$$(\sqrt{\sigma})'|_{y=0} = \sqrt{\sigma(y)}\, v(y,y)|_{y=0}.$$

The problem of recovery of $\sigma(y)$ from V and a given $q(y)$ is reduced to the Cauchy problem for the ordinary differential equation (1.6.12).

Another method of determination of the coefficients from V is based on the fact that the function

$$V^0(y) = 1 + \frac{1}{2}\int_{-y}^{y} v(y,t)\, \mathrm{d}t$$

is a solution of the differential equation

$$V^{0\prime\prime} + \Big(q(y) - \frac{(\sqrt{\sigma})''}{\sqrt{\sigma}}\Big)V^0 = 0. \tag{1.6.13}$$

Equation (1.6.13) is derived in a similar way as (1.4.11). Comparing formulas (1.6.12) and (1.6.13), we obtain the equality

$$\frac{V^{0\prime\prime}}{V^0} = 2\frac{\mathrm{d}}{\mathrm{d}y}v(y,y).$$

Therefore, either $V^0(y)$ or $v(y,y)$ can be used to find the desired coefficient. However, formula (1.6.12) is usually more convenient because the application of the function V^0 requires a) the additional integration of the solution of the Gel'fand – Levitan equation, b) the subsequent double differentiation and division by V^0: $V^{0\prime\prime}/V^0$ (note that V^0 may vanish). Nevertheless, in some cases it is more convenient to use V^0; for example, when $q \equiv 0$. In this case, (1.6.13) can be rewritten as

$$\Big(\Big(\frac{V^0}{\sqrt{\sigma}}\Big)'\sigma\Big)' = 0,$$

whence it is easily seen that

$$V^0(y) = \frac{\sqrt{\sigma(y)}}{\sqrt{\sigma(0)}}.$$

Thus, we can obtain $\sigma(y)$ explicitly from $V^0(y)$.

1.6.3. The scattering problem

The way of the study of the scattering problem is analogous to that used in Section 1.4 and Subsections 1.6.1, 1.6.2. Therefore, we shall restrict ourselves to a brief review.

Suppose that equation (1.6.1) holds for all $y \in (-\infty, \infty)$. For $y < 0$, we have $\sigma(y) \equiv 1$, $q(y) \equiv 0$. Let a function U solve equation (1.6.1) and, for $y < 0$,

$$U|_{y<0} = \delta(t-y) + f(t+y),$$
$$U|_{t<y} \equiv 0.$$

Parallel with U, we introduce a function W solving (1.6.1) subject to

$$W|_{y<0} = \delta(t-y).$$

It is easy to verify that

$$W|_{t>y} \equiv 0, \qquad W|_{t<-y,\, y>0} \equiv 0.$$

The formula

$$U = W(y,t) + \int_{-\infty}^{\infty} W(s,y)f(t+s)\,\mathrm{d}s \tag{1.6.14}$$

holds. The singularity of the function W is as follows:

$$W = \frac{1}{\sqrt{\sigma(y)}}(\delta(t-y) + w(y,t)).$$

Taking into account that $U|_{|t|<y} \equiv 0$ for $y > 0$, we obtain from (1.6.14), after reduction by $(\sigma(y))^{-1/2}$ the linear integral equation of second kind relative to the function w:

$$w(y,t) + \int_{-t}^{y} f(s+t)w(y,s)\,\mathrm{d}s + f(t+y) = 0. \tag{1.6.15}$$

The inverse problem is solvable iff, for any y, the equation

$$\varphi(t) + \int_{-t}^{y} f(s+t)\varphi(s)\,\mathrm{d}s = 0$$

has only the trivial solution. Recovery of one of the coefficients σ or q from the solution of (1.6.15) if the other is known can be realized by the scheme proposed in Subsection 1.6.2. By analogy with (1.6.12), we use the equality

$$2\frac{\mathrm{d}}{\mathrm{d}y}w(y,y) = \frac{(\sqrt{\sigma})''}{\sqrt{\sigma}} - q(y).$$

We can use also the function

$$W^0 = 1 + \int_{-y}^{y} w(y,t)\,\mathrm{d}t.$$

Remark 1.6.1. In the above considerations we have assumed that the function q is real. However, the case of nonreal q creates no additional difficulties.

1.7. SOME CASES OF EXPLICIT SOLUTION OF INVERSE PROBLEM

1.7.1. Description of inverse problem data

We consider equations (1.4.14) and (1.4.15). Note that the kernels of the integral operators entering these equations have the special structure: they depend on the sum or the difference of arguments. This implies that the integral equations (1.4.14), (1.4.15) have explicit solutions in some cases, namely, when the functions $h_1(t)$ and $h_2(t)$ are either finite linear combinations of exponentials or products of polynomials by exponentials. Besides, the exponentials may have complex powers. So, we assume that the functions h_1 and h_2 are finite sums of terms such as $a_{kj}t^k e^{\lambda_j t}$, where $k \geq 0$ is integer, a_{kj} and λ_j are complex numbers.[7]

Evidently, each linear combination of this type can be represented as

$$h_l(t) = \frac{1}{2\pi i} \int_{\mathcal{L}} e^{zt} \frac{P_l(z)}{Q(z)} \, dz. \tag{1.7.1}$$

Here $Q(z)$, $P_1(z)$, $P_2(z)$ are polynomials. The degrees of $P_1(z)$ and $P_2(z)$ are less than that of $Q(z)$. The polynomial $Q(z)$ has roots at the points $\lambda_1, \lambda_2, \ldots$, where λ_i are the constants entering the exponential powers as indicated above. Multiplicity of each root λ_i must exceed, at least by unity, the highest power k in the terms $t^k e^{\lambda_i t}$ which enter the linear combination. The integral in (1.7.1) is taken round a closed circuit $\mathcal{L}$ covering all zeros of the denominator. Really, calculating the integral by the residue theorem, we obtain the expression for $h_l(t)$ in the form of the required linear combination. It is easy to verify that varying $P_l(z)$ we can obtain any coefficients at the terms $t^k e^{\lambda_j t}$. The polynomial $Q(z)$ may be the same in the expressions for $h_1(t)$ and $h_2(t)$ because $Q(z)$ may have "excessive" roots. Indeed, let $h_1(t)$ and $h_2(t)$ have the terms $t^k e^{\lambda_0 t}$, with k_1 and k_2 being the highest powers of t among these terms in $h_1(t)$ and $h_2(t)$, respectively. Assuming $k_1 > k_2$ for definiteness, we obtain that λ_0 is the root of $Q(z)$ of multiplicity $k_1 + 1$. In this case, $P_2(\lambda_0) = 0$ and the fraction $P_2(z)/Q(z)$ is reducible by $(z - \lambda_0)^{k_1 - k_2}$. However, we assume that none of λ_j is the root of both the polynomials $P_l(z)$. This condition allows us to determine $Q(z)$ by $h_1(t)$, $h_2(t)$ uniquely to within a constant multiplier. The degree of $Q(z)$ is denoted by n.

1.7.2. Construction of the solution

To construct a solution we use the following scheme. Fix y and note that

[7] As a rule, only real functions $h_1(t)$, $h_2(t)$ are physically reasonable. Therefore, each term $at^k e^{\lambda t}$ in the linear combination with a nonreal λ necessitates the term $\bar{a}t^k e^{\bar{\lambda} t}$, where the bar denotes complex conjugation. Their sum has the form

$$\exp(\lambda' t) t^k (a_1 \cos \lambda'' t + a_2 \sin \lambda'' t)$$

with real coefficients a_1, a_2 and real parameters $\lambda', \lambda'' \, : \, \lambda' = \operatorname{Re} \lambda, \lambda'' = \operatorname{Im} \lambda$.

solutions of equations (1.4.14), (1.4.15) have the form

$$\chi_k(t) = \frac{1}{2\pi i}\int_{\mathcal{M}} e^{\zeta t}\frac{L_k(\zeta)}{S(\zeta)}\, d\zeta, \qquad k = 1, 2, \tag{1.7.2}$$

where

$$\begin{aligned} S(\zeta) &= \tilde{Q}(\zeta)\tilde{Q}(-\zeta) - P_2(\zeta)P_2(-\zeta), \\ L_k(\zeta) &= R_k(\zeta)\tilde{Q}(-\zeta) - (-1)^k R_k(-\zeta)P_2(\zeta), \\ \tilde{Q}(\zeta) &= Q(\zeta) - P_1(\zeta), \end{aligned}$$

$R_k(\zeta)$ are polynomials in ζ of degree $\leq n-1$ with arbitrary coefficients. The integral in (1.7.2) is taken round a circuit $\mathcal{M}$ on the complex plane of ζ such that a) all the roots of $S(\zeta)$ are inside $\mathcal{M}$; b) the contour $\mathcal{L}$ is inside $\mathcal{M}$; c) $\mathcal{M}$ is centrally symmetric (if $\zeta \in \mathcal{M}$, then $-\zeta \in \mathcal{M}$).

The operator $\hat{H}_1$ applied to $\chi_k(t)$ yields

$$\begin{aligned} \hat{H}_1\chi_k(t) &= \int_{-y}^{t} h_1(t-s)\chi_k(s)\, ds \\ &= \Big(\frac{1}{2\pi i}\Big)^2 \iint_{\mathcal{L}\times\mathcal{M}} dz\, d\zeta\, \frac{L_k(\zeta)}{S(\zeta)}\frac{P_1(z)}{Q(z)}\int_{-y}^{t} e^{z(t-s)}e^{\zeta s}\, ds \\ &= \Big(\frac{1}{2\pi i}\Big)^2 \iint_{\mathcal{L}\times\mathcal{M}} dz\, d\zeta\, \frac{L_k(\zeta)}{S(\zeta)}\frac{P_1(z)}{Q(z)}\frac{1}{\zeta - z}(e^{\zeta t} - e^{-\zeta y}e^{z(t+y)}) \\ &= \frac{1}{2\pi i}\int_{\mathcal{M}} d\zeta\, \frac{L_k(\zeta)}{S(\zeta)}\Big(e^{\zeta t}\frac{1}{2\pi i}\int_{\mathcal{L}}\frac{P_1(z)}{Q(z)}\frac{dz}{\zeta - z} \\ &\qquad - e^{-\zeta y}\frac{1}{2\pi i}\int_{\mathcal{L}}\frac{e^{z(t+y)}}{\zeta - z}\frac{P_1(z)}{Q(z)}\, dz\Big). \end{aligned}$$

Since

$$\frac{1}{2\pi i}\int_{\mathcal{L}}\frac{P_1(z)}{Q(z)}\frac{dz}{\zeta - z} = \frac{P_1(\zeta)}{Q(\zeta)},$$

we obtain

$$\begin{aligned} \hat{H}_1\chi_k(t) = \frac{1}{2\pi i}&\int_{\mathcal{M}} d\zeta\, \frac{L_k(\zeta)}{S(\zeta)}\frac{P_1(\zeta)}{Q(\zeta)}\, e^{\zeta t} \\ &- \frac{1}{2\pi i}\int_{\mathcal{L}}\frac{dz}{Q(z)}\, e^{z(t+y)}P_1(z)G_k^{(1)}(z) \end{aligned} \tag{1.7.3}$$

where $G_k^{(1)}(z)$ is a regular function of z inside the circuit $\mathcal{L}$:

$$G_k^{(1)}(z) = \frac{1}{2\pi \mathrm{i}} \int_{\mathcal{M}} \mathrm{d}\zeta \, \frac{L_k(\zeta)}{S(\zeta)} \frac{\mathrm{e}^{-\zeta y}}{\zeta - z}.$$

At the same time,

$$\begin{aligned}\hat{H}_2\chi_k(t) &= \int_{-t}^{y} h_2(t+s)\chi_k(s)\,\mathrm{d}s \\ &= \Big(\frac{1}{2\pi \mathrm{i}}\Big)^2 \iint_{\mathcal{L}\times\mathcal{M}} \mathrm{d}z\,\mathrm{d}\zeta \, \frac{L_k(\zeta)}{S(\zeta)} \frac{P_2(z)}{Q(z)} \int_{-t}^{y} \mathrm{e}^{z(t+s)}\mathrm{e}^{\zeta s}\,\mathrm{d}s.\end{aligned}$$

In this expression we change the integration variables: $\zeta \to -\zeta$ and $s \to -s$. Using the central symmetry of $\mathcal{M}$ and evenness of the polynomial $S(\zeta)$, we obtain

$$\hat{H}_2\chi_k(t) = -\Big(\frac{1}{2\pi \mathrm{i}}\Big)^2 \iint_{\mathcal{L}\times\mathcal{M}} \mathrm{d}z\,\mathrm{d}\zeta \, \frac{L_k(-\zeta)}{S(\zeta)} \frac{P_2(z)}{Q(z)} \int_{-y}^{t} \mathrm{e}^{z(t-s)}\mathrm{e}^{\zeta s}\,\mathrm{d}s.$$

The representation of $\hat{H}_2\chi_k(t)$ differs from that of $\hat{H}_1\chi_k(t)$ in 1) the sign; 2) the sign of the argument of L_k; 3) $P_2(z)$ instead of $P_1(z)$. Applying transformations similar to those used for $\hat{H}_1\chi_k(t)$, we arrive at

$$\hat{H}_2\chi_k(t) = -\frac{1}{2\pi \mathrm{i}} \int_{\mathcal{M}} \mathrm{d}\zeta \, \frac{L_k(-\zeta)}{S(\zeta)} \frac{P_2(\zeta)}{Q(\zeta)} \mathrm{e}^{\zeta t} + \frac{1}{2\pi \mathrm{i}} \int_{\mathcal{L}} \frac{\mathrm{d}z}{Q(z)} \, \mathrm{e}^{z(t+y)} P_2(z) G_k^{(2)}(z),$$

$$G_k^{(2)}(z) = \frac{1}{2\pi \mathrm{i}} \int_{\mathcal{M}} \mathrm{d}\zeta \, \frac{L_k(-\zeta)}{S(\zeta)} \frac{\mathrm{e}^{-\zeta y}}{\zeta - z}. \tag{1.7.4}$$

Substituting this representation into equations (1.4.14), (1.4.15), we obtain

$$0 = \chi_k(t) + \hat{H}_1\chi_k(t) + (-1)^{k-1}(\hat{H}_2\chi_k(t) + h_2(t+y)) = \Phi_k(t) + \Psi_k(t)$$

where

$$\begin{aligned}\Phi_k(t) &= \frac{1}{2\pi \mathrm{i}} \int_{\mathcal{M}} \mathrm{d}\zeta \, \frac{\mathrm{e}^{\zeta t}}{Q(\zeta)S(\zeta)} \Big[(Q(\zeta) + P_1(\zeta))L_k(\zeta) + (-1)^k P_2(\zeta) L_k(-\zeta)\Big] \\ &= \frac{1}{2\pi \mathrm{i}} \int_{\mathcal{M}} \mathrm{d}\zeta \, \frac{\mathrm{e}^{\zeta t}}{Q(\zeta)S(\zeta)} \Big[\tilde{Q}(\zeta)(R_k(\zeta)\tilde{Q}(-\zeta) - (-1)^k R_k(-\zeta)P_2(\zeta)) \\ &\qquad + (-1)^k P_2(\zeta)(R_k(-\zeta)\tilde{Q}(\zeta) - (-1)^k R_k(\zeta)P_2(-\zeta))\Big]\end{aligned}$$

$$= \frac{1}{2\pi \mathrm{i}} \int_{\mathcal{M}} \mathrm{d}\zeta \, \frac{\mathrm{e}^{-\zeta t}}{Q(\zeta)S(\zeta)} \left[R_k(\zeta)(\tilde{Q}(\zeta)\tilde{Q}(-\zeta) - P_2(\zeta)P_2(-\zeta)) \right]$$

$$= \frac{1}{2\pi \mathrm{i}} \int_{\mathcal{M}} \mathrm{d}\zeta \, \frac{\mathrm{e}^{-\zeta t}}{Q(\zeta)} R_k(\zeta) = \frac{1}{2\pi \mathrm{i}} \int_{\mathcal{L}} \mathrm{d}z \, \frac{\mathrm{e}^{zt}}{Q(z)} R_k(z),$$

$$\Psi_k(t) = \frac{1}{2\pi \mathrm{i}} \int_{\mathcal{L}} \mathrm{d}z \, \frac{\mathrm{e}^{z(t+y)}}{Q(z)} \left[P_1(z) G_k^{(1)}(z) + (-1)^{k-1} P_2(z)(G_k^{(2)}(z) + 1) \right].$$

The expression $\Phi_k(t) + \Psi_k(t)$ is equal to zero if and only if the roots of the function

$$R_k(z) + \mathrm{e}^{zy} P_1(z) G_k^{(1)}(z) + (-1)^{k-1} \mathrm{e}^{zy} P_2(z)(G_k^{(2)}(z) + 1)$$

and their multiplicities coincide with those of $Q(z)$. Since $G_k^{(1)}(z)$ and $G_k^{(2)}(z)$ are linearly expressed in terms of the polynomials $R_k(z)$, the last requirement is reduced to a linear algebraic system of n equations relative to n unknowns (the coefficients of the polynomial $R_k(z)$). All the functions involved in this construction are elementary (polynomials, exponentials).

1.8. ON CONNECTION OF INVERSE PROBLEMS WITH NONLINEAR ORDINARY DIFFERENTIAL EQUATIONS

1.8.1. Connection between ordinary differential equations and inverse problems

As was shown in (Gardner *et al.*, 1967), there exists a remarkable connection between inverse scattering problems and nonlinear differential equations. This connection allows us, in particular, to construct explicit solutions of nonlinear problems. In what follows, we shall establish another connection between inverse problems and nonlinear differential equations which differs essentially from that mentioned in (Gardner *et al.*, 1967).

As noted above (see Remark 1.3.2), the inverse problem (1.1.12)–(1.1.14) is equivalent to the Cauchy problem for the nonlinear hyperbolic system with the shifted argument:

$$\vec{\Psi}_t + \Lambda \vec{\Psi}_y = \Psi_2(y, y) \begin{pmatrix} 0 & 1 \\ -1 & 0 \end{pmatrix} \vec{\Psi}, \qquad t > y > 0, \tag{1.8.1}$$

$$\vec{\Psi}\big|_{y=0} = \vec{h}(t). \tag{1.8.2}$$

Assume that $\vec{h}(t)$ has the form $\vec{h}(t) = (h_1(t), h_2(t))$ where

$$h_i(t) = \sum_{k=1}^{m} e^{\lambda_k t} \sum_{j=0}^{r_k} \alpha_{jk}^{(i)} \frac{t^j}{j!}, \qquad t > y. \tag{1.8.3}$$

Here λ_k and $\alpha_{jk}^{(i)}$ are complex constants. It is easy to verify that the functions $\Psi_i(y,t)$, for all $y > 0$, have a similar structure:

$$\Psi_i(y,t) = \sum_{k=1}^{m} e^{\lambda_k (t-y)} \sum_{j=0}^{r_k} x_{jk}^{(i)}(y) \frac{(t-y)^j}{j!}, \qquad t > y. \tag{1.8.4}$$

We introduce a pair of vector functions

$$X^{(i)}(y) \in \mathbb{C}^n \qquad (i = 1, 2),$$

where $n = \sum_{k=1}^{m} (r_k + 1)$, as follows:

$$X^{(i)}(y) = (x_{01}^{(i)}, x_{11}^{(i)}, \dots, x_{r_1 1}^{(i)}, \dots, x_{0m}^{(i)}, \dots, x_{r_m m}^{(i)}).$$

Substituting (1.8.4) in (1.8.1), (1.8.2), we obtain that the pair of the functions $\Psi_1(y,t)$, $\Psi_2(y,t)$ of the form (1.8.4) solves the problem (1.8.1), (1.8.2) if $X^{(1)}(y)$ and $X^{(2)}(y)$ are solutions of the system

$$\frac{d}{dy} X^{(1)} = l(X^{(2)}) X^{(2)},$$

$$\frac{d}{dy} X^{(2)} = l(X^{(2)}) X^{(1)} + 2C X^{(2)}, \tag{1.8.5}$$

subject to the initial condition

$$X^{(i)}\big|_{y=0} = X^{(i)0}.$$

Here C is a quasidiagonal matrix consisting of Jordan boxes:

$$C = \begin{pmatrix} C_1 & 0 & . & 0 \\ 0 & C_2 & . & 0 \\ . & . & . & . \\ 0 & 0 & . & C_m \end{pmatrix}, \qquad C_k = \begin{pmatrix} \lambda_k & 1 & 0 & . & 0 & 0 \\ 0 & \lambda_k & 1 & . & 0 & 0 \\ . & . & . & . & . & . \\ 0 & 0 & 0 & . & \lambda_k & 1 \\ 0 & 0 & 0 & . & 0 & \lambda_k \end{pmatrix},$$

$l(X^{(2)})$ is a linear combination of the form

$$l(X^{(2)}) = \sum_{k=1}^{m} x_{0k}^{(2)}(y) = \Psi_2(y,y),$$

$$X^{(i)0} = (\alpha_{01}^{(i)}, \alpha_{11}^{(i)}, \dots, \alpha_{r_1 1}^{(i)}, \dots, \alpha_{0m}^{(i)}, \dots, \alpha_{r_m m}^{(i)}).$$

1.8.2. Use of the connection between differential equations and inverse problems

The connection under consideration can be used in two directions.

1. Since the inverse problem with data (1.8.3) is solvable explicitly, this can be used for constructing an explicit solution of the Cauchy problem for the system of differential equations (1.8.5). Note that the above characterization of the matrix C and the linear form l is inessential except for the property that $l(Y^0) \neq 0$ for each eigenvector Y^0 of the matrix C. It can be shown that if this property holds, then using a linear change of variables, l and C can be reduced to the special form described above.

2. Conversely, even though the data has the form (1.8.3), we may face serious computational difficulties in applying explicit formulas to the inverse problem. In this case, it is recommended to solve, instead of the inverse problem, the Cauchy problem for system (1.8.5) using known numerical methods. The subsequent discussion follows the paper (Blagoveshchenskii, 1995).

As an example, we consider the string equation

$$u_{tt} + \sigma u_t = c^2 u_{xx} + a u_x + bu, \tag{1.8.6}$$

$x \in (0, l)$ $(0 < l \leq \infty)$; σ, a, b, c are continuous (in general, complex-valued) functions of the variable x; $c = c(x) > 0$. Without loss of generality, we may take $c(x) \equiv 1$. This equality if false can be attained by changing the variable $x' = \int_0^x ds/c(s)$. Assume that the wave process in the medium is generated by a source localized at the point $x = 0$. The source begins to act at the moment $t = 0$ and generates the wave $u(x, t)$ which has a jump on its leading edge. So, we assume that the wave satisfies the condition

$$u(x,t)|_{t<x} = 0,$$

and $u(x, t)$ is smooth for $x < t < 2l - x$.

We shall consider the problems of finding one of the functions σ, a, b (the other two are supposed as known) from the given functions $f(t)$, $g(t)$:

$$f(t) := u(x,t)|_{x=0}, \tag{1.8.7}$$

$$g(t) := \frac{\partial u(x,t)}{\partial x}\Big|_{x=0} + \delta(t). \tag{1.8.8}$$

Here δ is the Dirac function, $f(t) = g(t) = 0$ for $t < 0$, $f \in C^1_{\mathrm{loc}}$, $g \in C_{\mathrm{loc}}$, $f(+0) = 1$.

Behavior of the solution $u(x,t)$ in a neighborhood of the characteristic $t = x$ is well known. Let

$$k(x) := u(x,t)|_{t=x+0}, \tag{1.8.9}$$

$$l(x) := \frac{\partial u(x,t)}{\partial t}\Big|_{t=x+0}. \tag{1.8.10}$$

Lemma 1.8.1. *If $\sigma, a, b \in C_{\mathrm{loc}}$, then*

$$\frac{\mathrm{d}k}{\mathrm{d}x} = -\frac{\sigma + a}{2}k, \tag{1.8.11}$$

$$k(x) = \exp\Big[-\frac{1}{2}\int_0^x (\sigma(\xi) + a(\xi))\,\mathrm{d}\xi\Big]. \tag{1.8.12}$$

In the case when $\sigma, a \in C^1_{\mathrm{loc}}$, $b \in C_{\mathrm{loc}}$, there holds

$$\frac{\mathrm{d}l}{\mathrm{d}x} = -\frac{\sigma + a}{2}l + \frac{1}{2}(k'' + ak' + bk). \tag{1.8.13}$$

The proof of this lemma is similar to that outlined in Subsection 1.2.3.

Change now from t to a new variable $s = t - x$ which, for $t > x$, models the elapsed time since the first appearance of the wave at the point x. Define the function

$$v(x,s) := u(x, x+s).$$

We shall consider the following three inverse problems:

Problem I is to determine σ;

Problem II is to determine a;

Problem III is to determine b.

Recall that in these inverse problems the two remaining coefficients of equation (1.8.6) are assumed as known.

Lemma 1.8.1 leads to the following statement.

Lemma 1.8.2. *Problems I, II, III are equivalent to the Cauchy problems with the initial data*

$$v|_{x=0} = f(s), \qquad \frac{\partial v(x,s)}{\partial x}\Big|_{x=0} = g(s)$$

in the domain $0 < x < l$, $0 < s < 2l - x$ *for the following equations*
(i) (for Problem I)

$$v_{xx} - 2v_{xs} = -\frac{2k'}{k} v_s - a v_x - b v, \tag{1.8.14}$$

(ii) (for Problem II)

$$v_{xx} - 2v_{xs} = -\frac{2k'}{k} v_s + \Big(\frac{2k'}{k} + \sigma\Big) v_x - b v, \tag{1.8.15}$$

(iii) (for Problem III)

$$\begin{aligned} v_{xx} - 2v_{xs} = {} & (\sigma + a) v_s - a v_x \\ & - \Big[\exp\Big(\frac{1}{2}\int_0^x (a+\sigma)\,\mathrm{d}\xi\Big)(2l' + (a+\sigma)l) \\ & + \frac{(a+\sigma)'}{2} + \frac{a^2 - \sigma^2}{4}\Big] v, \end{aligned} \tag{1.8.16}$$

with $k = v(x, +0)$, $l = (\partial v/\partial s)(x, +0)$.

Equations (1.8.14)–(1.8.16) are nonlinear hyperbolic equations with the shifted argument.

For Problem I or Problem II to be solved, we can find a solution of equation (1.8.14) or (1.8.15), respectively, determine the function k, and then recover the function σ or a using relation (1.8.11).

To solve Problem III, we determine a solution of equation (1.8.16), find l and k, and then recover the function b by formula (1.8.13).

Lemma 1.8.3. *Equations* (1.8.14)–(1.8.16) *have particular solutions of the form*

$$v(x,s) = \sum_{k=1}^{m} \sum_{j=o}^{n_k} y_{kj}(x)\, \frac{s^j}{j!}\, \mathrm{e}^{\lambda_k s} \tag{1.8.17}$$

with mutually distinct λ_k *and integer* $m \geq 1$, $n_1 \geq 0, \dots, n_m \geq 0$.

The lemma is proved by substitution of (1.8.17) into equations (1.8.14)–(1.8.16).

The function (1.8.17) solves equation (1.8.14) if the set of functions y_{kj} satisfies the following system of ordinary differential equations:

$$y''_{kj} = 2y'_{k\,j+1} + 2y'_{kj}\lambda_k - 2\,\frac{\sum_{p=1}^{m} y'_{p0}}{\sum_{p=1}^{m} y_{p0}}\,(y_{k\,j+1} + \lambda_k y_{kj}) - a(x) y'_{kj} - b(x) y_{kj}. \tag{1.8.18}$$

If y_{kj} satisfy the system

$$y''_{kj} = 2y'_{k\,j+1} + 2y'_{kj}\lambda_k - 2\,\frac{\sum_{p=1}^m y'_{p0}}{\sum_{p=1}^m y_{p0}}\,(y_{k\,j+1} + \lambda_k y_{kj} - y'_{kj}) + \sigma y'_{kj} - b(x)y_{kj}, \tag{1.8.19}$$

then the function (1.8.17) solves equation (1.8.15).

If y_{kj} satisfy the system

$$\begin{aligned} y''_{kj} &= 2y'_{k\,j+1} + 2y'_{kj}\lambda_k + (\sigma + a)(y_{k\,j+1} + \lambda_k y_{kj}) - a y'_{kj} \\ &- \Big[2\sum_{p=1}^m \Big(y'_{p1} + \lambda_p y'_{p0} + \frac{a(x)+\sigma(x)}{2}\,(y_{p1} + \lambda_p y_{p0})\Big) \\ &\times \exp\Big(\frac{1}{2}\int_0^x (a+\sigma)\,\mathrm{d}x\Big) + \frac{(a+\sigma)'}{2} + \frac{a^2-\sigma^2}{4}\Big] y_{kj}, \end{aligned} \tag{1.8.20}$$

then the function (1.8.17) solves equation (1.8.16).

Evidently, the solution $v(x,s)$ has the form (1.8.17) if the initial data are similarly representable:

$$v(x,s)\big|_{x=0} = f(s) = \sum_{k=1}^m \sum_{j=0}^{n_k} \alpha_{kj}\,\frac{s^j}{j!}\,\mathrm{e}^{\lambda_k s}, \tag{1.8.21}$$

$$\frac{\partial v(x,s)}{\partial x}\bigg|_{x=0} = g(s) = \sum_{k=1}^m \sum_{j=0}^{n_k} \beta_{kj}\,\frac{s^j}{j!}\,\mathrm{e}^{\lambda_k s}. \tag{1.8.22}$$

Thus, in the case when the initial data have the form (1.8.21), (1.8.22), the problems in question can be solved as follows: first we solve the Cauchy problem for the system of ordinary differential equations (1.8.18), (1.8.19), or (1.8.20) with the initial data

$$y_{kj}\big|_{x=0} = \alpha_{kj}, \qquad y'_{kj}\big|_{x=0} = \beta_{kj}; \tag{1.8.23}$$

then we construct the functions k and l

$$k = v(x,+0) = \sum_{k=1}^m y_{k0}(x),$$

$$l = \frac{\partial v}{\partial s}(x,+0) = \sum_{k=1}^m (\lambda_k y_{k0}(x) + y_{k1}(x));$$

and finally we recover the desired function.

1.9. THE CASE OF MORE GENERAL SYSTEM OF EQUATIONS

1.9.1. Setting of a problem

We shall consider here another inverse problem (see [41]). The inverse problem seems to be unnatural from the physical point of view. However, in Subsection 2.3.1, we shall demonstrate some advantages of this setting.

Consider, for $y > 0$, the system of equations

$$\begin{aligned} \Psi_{1t} + \Psi_{1y} &= q_1 \Psi_2, \\ \Psi_{2t} - \Psi_{2y} &= q_2 \Psi_1 \end{aligned} \tag{1.9.1}$$

where q_1, q_2 are functions of y. Denote by $\vec{\Psi}^+$ the solution of (1.9.1) which satisfies the conditions

$$\vec{\Psi}^+\big|_{t<y} = 0, \qquad \vec{\Psi}^+\big|_{y=0} = \vec{H}^+(t) = \begin{pmatrix} \delta(t) + h_1^+(t) \\ h_2^+(t) \end{pmatrix}. \tag{1.9.2}$$

The functions $h_i^+(t)$, as well as $h_i^-(t)$ introduced below, are piecewise continuous. The solution $\vec{\Psi}^+$ has evident physical meaning. Along with this solution, we consider the solution $\vec{\Psi}^-$ of (1.9.1) subject to the conditions

$$\vec{\Psi}^-\big|_{t>-y} = 0, \qquad \vec{\Psi}^-\big|_{y=0} = \vec{H}^-(t) = \begin{pmatrix} h_1^-(t) \\ \delta(t) + h_2^-(t) \end{pmatrix}. \tag{1.9.3}$$

The solution $\vec{\Psi}^-$ models the wave field which is completely quenched by the time $t = 0$ by a singular wave coming from the right. This solution makes few physical sense. Nevertheless, it exists. Indeed, if we change $t \to -t$ and interchange the roles of the components of the vector function $\vec{\Psi}^-$, the solution will be of the type $\vec{\Psi}^+$ for system (1.9.1) with the changed coefficients: q_1 is replaced by $-q_2$; and q_2, by $-q_1$.

Note that when $q_1(y)$ and $q_2(y)$ are fixed, the functions $h_1^+(t)$ and $h_2^+(t)$ are not independent. For the unique determination of $\vec{\Psi}^+$ it suffices to know $h_1^+(t)$. The same holds for the functions $h_1^-(t)$, $h_2^-(t)$. Therefore, given both the pairs $h_i^{\pm}(t)$, we have overdetermined information, which enables us to set the inverse problem of determination of both the functions q_1, q_2.

1.9.2. Linear integral equations

We consider another pair of solutions $\vec{\Theta}^{\pm}(y,t)$ of system (1.9.1) which meet the conditions

$$\vec{\Theta}^{+}(y,t)\big|_{y=0} = \begin{pmatrix} \delta(t) \\ 0 \end{pmatrix}, \qquad \vec{\Theta}^{-}(y,t)\big|_{y=0} = \begin{pmatrix} 0 \\ \delta(t) \end{pmatrix}.$$

These solutions have the following structure:

$$\vec{\Theta}^{+}(y,t) = \begin{pmatrix} \delta(t-y) \\ 0 \end{pmatrix} + \vec{\theta}^{+}(y,t), \qquad \vec{\Theta}^{-}(y,t) = \begin{pmatrix} 0 \\ \delta(t+y) \end{pmatrix} + \vec{\theta}^{-}(y,t).$$

Here $\vec{\theta}^{\pm}$ are piecewise continuous functions which differ from zero only for $t \in (-y,y)$. As before, from $\vec{\Theta}^{\pm}(y,t)$ we can construct each solution and, in particular, $\vec{\Psi}^{\pm}(y,t)$:

$$\begin{aligned} \vec{\Psi}^{+}(y,t) &= \vec{\Theta}^{+}(y,t) * H_1^{+}(t) + \vec{\Theta}^{-}(y,t) * H_2^{+}(t) \\ &= \vec{\Theta}^{+}(y,t) + \vec{\Theta}^{+}(y,t) * h_1^{+}(t) + \vec{\Theta}^{-}(y,t) * h_2^{+}(t), \\ \vec{\Psi}^{-}(y,t) &= \vec{\Theta}^{+}(y,t) * H_1^{-}(t) + \vec{\Theta}^{-}(y,t) * H_2^{-}(t) \\ &= \vec{\Theta}^{-}(y,t) + \vec{\Theta}^{+}(y,t) * h_1^{-}(t) + \vec{\Theta}^{-}(y,t) * h_2^{-}(t). \end{aligned} \tag{1.9.4}$$

Since $\vec{\Psi}^{\pm}(y,t) = 0$ for $t \in (-y,y)$, we obtain the system of four scalar (two vector) integral equations relative to $\theta_i^{\pm}$ $(i = 1,2)$:

$$\begin{aligned} &\theta_1^{+}(y,t) + \int_{-y}^{t} h_1^{+}(t-t_1)\theta_1^{+}(y,t_1)\,\mathrm{d}t_1 + \int_{-y}^{t} h_2^{+}(t-t_1)\theta_1^{-}(y,t_1)\,\mathrm{d}t_1 = 0, \\ &\theta_1^{-}(y,t) + \int_{t}^{y} h_1^{-}(t-t_1)\theta_1^{+}(y,t_1)\,\mathrm{d}t_1 \\ &\qquad + \int_{t}^{y} h_2^{-}(t-t_1)\theta_1^{-}(y,t_1)\,\mathrm{d}t_1 + h_1^{-}(t-y) = 0, \\ &\theta_2^{+}(y,t) + \int_{-y}^{t} h_1^{+}(t-t_1)\theta_2^{+}(y,t_1)\,\mathrm{d}t_1 \\ &\qquad + \int_{-y}^{t} h_2^{+}(t-t_1)\theta_2^{-}(y,t_1)\,\mathrm{d}t_1 + h_2^{+}(t+y) = 0, \\ &\theta_2^{-}(y,t) + \int_{t}^{y} h_1^{-}(t-t_1)\theta_2^{+}(y,t_1)\,\mathrm{d}t_1 + \int_{t}^{y} h_2^{-}(t-t_1)\theta_2^{-}(y,t_1)\,\mathrm{d}t_1 = 0. \end{aligned}$$

Here we have written firstly the first components of the vector equalities (1.9.4) and then the second ones. The integrated terms $h_1^{-}(t-y)$

and $h_2^+(t+y)$ in the second and third equations have appeared because of δ-singularity in the first component of $\vec{\theta}^+(y,t)$ and the second component of $\vec{\theta}^-(y,t)$. We see that the system of four integral equations falls into two systems relative to the pairs of functions $\theta_1^+(y,t)$, $\theta_1^-(y,t)$ and $\theta_2^+(y,t)$, $\theta_2^-(y,t)$.

Introducing the integral operators

$$\hat{H}_i^+ = \int_{-y}^{t} h_i^+(t-t_1)(\cdot)\,\mathrm{d}t_1, \qquad \hat{H}_i^- = \int_{-y}^{t} h_i^-(t-t_1)(\cdot)\,\mathrm{d}t_1 \qquad (i=1,2),$$

we can write these systems as follows:

$$\begin{aligned} &\theta_1^+ + \hat{H}_1^+\theta_1^+ + \hat{H}_2^+\theta_1^- = 0,\\ &\theta_1^- + \hat{H}_1^-\theta_1^+ + \hat{H}_2^-\theta_1^- + h_1^-(t-y) = 0, \end{aligned} \tag{1.9.5}$$

$$\begin{aligned} &\theta_2^+ + \hat{H}_1^+\theta_2^+ + \hat{H}_2^+\theta_2^- + h_2^+(t+y) = 0,\\ &\theta_2^- + \hat{H}_1^-\theta_2^+ + \hat{H}_2^-\theta_2^- = 0. \end{aligned} \tag{1.9.6}$$

Each of the integral operators $\hat{H}_i^\pm$ involved in these systems is of Volterra type. However, $\hat{H}_i^+$ and $\hat{H}_i^-$ are Volterra operators "in different directions". Therefore, either of the systems is a Fredholm system of second kind. Systems (1.9.5), (1.9.6) can be written briefly as

$$(E+K_y)\Theta_i + H^{(i)} = 0$$

where Θ_i is the vector function $(\theta_i^+(y,t), \theta_i^-(y,t))$, $i=1,2$; $H^{(i)}$ is equal to $(0, h_1^-(t-y))$ for $i=1$ and $(h_2^+(t+y), 0)$ for $i=2$; K_y is a matrix integral operator.

1.9.3. Determination of q_1 and q_2

To solve the inverse problem, it suffices to solve systems (1.9.5), (1.9.6). Really, it is easy to see that the functions $\vec{\theta}^\pm(y,t)$ have the following singularities (we restrict ourselves to the principal terms):

$$\begin{aligned} \theta_1^+(y,t) &= \delta(y-t) - \frac{1}{2}\int_0^y q_1(\xi)q_2(\xi)\,\mathrm{d}xi\,\varepsilon(y-t) + \dots,\\ \theta_1^-(y,t) &= \frac{q_1}{2}\,\varepsilon(t+y) + \dots,\\ \theta_2^+(y,t) &= -\frac{q_2}{2}\,\varepsilon(y-t) + \dots,\\ \theta_2^-(y,t) &= \delta(t+y) - \frac{1}{2}\int_0^y q_1(\xi)q_2(\xi)\,\mathrm{d}xi\,\varepsilon(y+t) + \dots. \end{aligned}$$

Thus,

$$\theta_1^-|_{t=-y+0} = \frac{q_1}{2}, \qquad \theta_2^+|_{t=y-0} = -\frac{q_2}{2},$$

which allows us to recover the coefficients of system (1.9.1) from (θ_1^+, θ_1^-) and (θ_2^+, θ_2^-). If we know the solution of one of the systems, say, (θ_1^+, θ_1^-), we can find

$$q_1 = 2\theta_1^-|_{t=-y+0}.$$

In the case when q_1 does not vanish, q_2 can easily be found from the value of jump of $\theta_1^+(y,t)$ at $t = y$:

$$\theta_1^+(y, y-0) = -\frac{1}{2}\int_0^y q_1(\xi)q_2(\xi)\,\mathrm{d}\xi.$$

Hence,

$$q_2(y) = -\frac{2}{q_1}\frac{\mathrm{d}}{\mathrm{d}y}\theta_1^+(y, y-0).$$

The necessary and sufficient condition of solvability of the inverse problem (1.9.1)–(1.9.3) is invertibility of the operator $(E + K_y)$. We shall not give here the proof of this fact because it is similar to that in Section 1.4. We shall only outline the proof of necessity. Suppose that

$$P(t) = \begin{pmatrix} p^+(t) \\ p^-(t) \end{pmatrix}$$

belongs to the kernel of the operator $(E + K_{y_0}^*)$ where $K_{y_0}^*$ is conjugate to K_{y_0}. We assume that $p^\pm(t)$ vanishes outside the interval $(-y_0, y_0)$. The proof is based on the solution of system (1.9.1):

$$(p^+(-t)) * \Psi^+ + (p^-(-t)) * \Psi^-.$$

On checking that this solution is equal to zero in the domain $\{y, t \mid |t| < y_0 - y\}$ we can obtain the homogeneous Volterra integral equations of second kind relative to $p^+(t)$ and $p^-(t)$, whence $p^+(t) \equiv p^-(t) \equiv 0$.

Chapter 2.

Theory of inverse problems for wave processes in layered media

2.1. INVERSE PROBLEMS OF ACOUSTICS

2.1.1. General remarks

Propagation of nonstationary waves in layered media is described, as a rule, by hyperbolic equations or systems of equations whose coefficients are functions of one coordinate though the modelled processes are multidimensional. This chapter is devoted to scalar processes; the next chapter deals with vector processes.

We shall consider here the inverse problems of finding the coefficients of the equations from the measurements of the wave fields on the medium boundary. Since the desired coefficients are dependent only on one coordinate, these problems are often called one-dimensional. We use here another terminology. We shall say that inverse problems are one-dimensional if they are similar to the problems considered in Chapter 1; i. e., if the wave field depends only on one spatial variable. The problems for layered media have a specificity which will be described below.

Since the coefficients are independent of some coordinates, we can apply to the direct problem the separation of variables, in one form or another. This generates a set of inverse problems; the role of parameter is played by the parameter of the separation of variables. The problems of this set are one-dimensional in the sense mentioned above. In this case the desired coefficients are usually functions of the parameter. Therefore, giving different values to the parameter, we can collect a great body of information (sometimes complete) on the coefficients of the initial equation.

Summarizing, we note that the investigation of inverse problems in layered media has two stages: 1) solving one-dimensional inverse problems; 2) finding maximal possible information on properties of the medium by varying the parameters. In some problems, both the stages are trivial (the first stage is regarded as trivial if it can be solved by the methods considered in the first chapter). However, there exist problems with one nontrivial stage, or both.

2.1.2. Setting of an inverse problem for the acoustic equation

In the half-space $z > 0$ with the Cartesian coordinates x_1, x_2, z, we consider the differential equation

$$a\frac{\partial^2 u}{\partial t^2} = \operatorname{div}(b \operatorname{grad} u). \tag{2.1.1}$$

Equation (2.1.1) arises when describing wave processes in physics, in particular, the sound propagation (in this case u is the pressure, $a = 1/\rho c^2$, $b = 1/\rho$, ρ is the medium density, c is the sound speed). Suppose that the function $u(x, z, t)$, $x = (x_1, x_2)$, satisfies the initial and boundary conditions

$$u|_{t<0} = 0, \tag{2.1.2}$$

$$bu_z|_{z=0} = \delta(x, t). \tag{2.1.3}$$

Coefficients a and b are assumed to be positive and dependent only on the coordinate z (stratification of the medium).

The problem (2.1.1)–(2.1.3) of determination of $u(x, z, t)$ if a and b are known is a well-posed direct problem of mathematical physics. Therefore, if we impose additional conditions on the function $u(x, z, t)$, they hold not for each $a(z)$ and $b(z)$: these conditions are equivalent to certain restrictions imposed on a and b. If additional conditions are numerous, these restrictions may be so rigid that a and b are determined uniquely (these conditions may even lead to inconsistency). Suppose the additional condition (the inverse problem data) to be given by

$$u|_{z=0} = f(x, t). \tag{2.1.4}$$

It is natural to expect that the information provided by the function $f(x,t)$ is excessive. Really, we need to determine two functions of one variable z from one function of three variables[1] $x = (x_1, x_2)$ and t; or, in other words, from a two-dimensional continuum of functions of one variable. Therefore, we may expect that, first, the solution of the inverse problem (2.1.1)–(2.1.4) is unique; second, for its existence the function $f(x,t)$ must satisfy rather rigid requirements. These expectations will turn out to be met. One of the restrictions on the function f is evident: the function $f(x,t)$, really, is a function of two (not three) variables. Indeed, the direct problem is axially symmetric. Therefore, its solution u must have the similar property:

$$u(x,z,t) = u(|x|,z,t), \qquad |x| = \sqrt{x_1^2 + x_2^2}.$$

Hence, $f(x,t) = u|_{z=0}$ must depend only on $|x|$ and t.

2.1.3. Solving the inverse problem of acoustics by the Fourier transformation

We apply to (2.1.1)–(2.1.4) the Fourier transformation with respect to the variable x:

$$\tilde{u}(k,z,t) = \int \mathrm{e}^{\mathrm{i}(k,x)} u(x,z,t)\,\mathrm{d}x$$

where $k = (k_1, k_2)$; $(k,x) = k_1x_1 + k_2x_2$; $\int \ldots \mathrm{d}x$ denotes the double integral taken over the whole plane of x_1, x_2. Note that in the processes described by equation (2.1.1), the propagation velocity is finite and equal to $\sqrt{b/a}$. Therefore, for each z and t we integrate in fact over a bounded domain outside of which $u \equiv 0$. The function $\tilde{u}(k,z,t)$ evidently satisfies the following conditions:

$$a\tilde{u}_{tt} = \frac{\partial}{\partial z}\Big(b\frac{\partial \tilde{u}}{\partial z}\Big) - b|k|^2\tilde{u}, \tag{2.1.5}$$

$$\tilde{u}|_{t<0} = 0, \tag{2.1.6}$$

$$b\tilde{u}_z|_{z=0} = \delta(t), \tag{2.1.7}$$

$$\tilde{u}|_{z=0} = \tilde{f}(|k|,t), \tag{2.1.8}$$

where $\tilde{f}(|k|,t)$ is the Fourier transform of $f(x,t)$. (It is easy to see that $\tilde{f}$ as a function of k depends on $|k| = \sqrt{k_1^2 + k_2^2}$.) Let $k = 0$. Then the problem (2.1.5)–(2.1.8) can be so rewritten that it coincides with the inverse

[1]These considerations can not be regarded as a proof. Nevertheless they often lead to true results.

problem (1.1.1)–(1.1.4) considered in Chapter 1. Solving this problem, we find $\sigma(y) = \sqrt{a(y)\, b(y)}$ where

$$y = \int_0^z \sqrt{\frac{a}{b}}\, \mathrm{d}z. \tag{2.1.9}$$

We choose now an arbitrary $k \neq 0$. Using (2.1.9), we change from z to the coordinate y and see that equation (2.1.5) becomes as follows:

$$\tilde{u}_{tt} = \frac{1}{\sigma(y)} \frac{\partial}{\partial y}(\sigma \tilde{u}_y) - k^2 \frac{b}{a} \tilde{u}. \tag{2.1.10}$$

This equation coincides with equation (1.6.1) if we set

$$q(y) = -k^2 \frac{b}{a} = -k^2 c^2.$$

Transformation of conditions (2.1.6)–(2.1.8) when changing to y is evident. The inverse problem (2.1.10), (2.1.6)–(2.1.8) coincides with the inverse problem considered in Section 1.6. The function $\sigma(y)$ can be thought of as known: it is the same as above. As in Section 1.6, we find $q(y)$. From $q(y) = -k^2 c^2$ we find $c(y)$. Finally, using the relation

$$y'_z = \sqrt{\frac{a}{b}} = \frac{1}{c(y)},$$

we can express z in terms of y:

$$z = \int_0^y c(y)\, \mathrm{d}y.$$

If we know the connection between z and y, the functions $c(y)$ and $\sigma(y)$, we can find $c(z)$, $\sigma(z)$ and therefore $a = \sigma / c$, $b = \sigma c$.

Thus, to find uniquely the coefficients a and b of equation (2.1.1), it suffices to know the values of the function $\tilde{f}(|k|, t)$ for two values of $|k|$: $|k| = 0$ and an arbitrary $|k| > 0$. We can also solve the inverse problem if we know $\tilde{f}(|k|, t)$ for two different nonzero values of $|k|$. Really, solving the inverse problems (2.1.10), (2.1.6)–(2.1.8) for $|k| = |k^{(1)}|$ and $|k| = |k^{(2)}|$, we find, as in Section 1.6, the functions

$$q_1(y) = -|k^{(1)}|^2 c^2(y) - \frac{(\sigma)''}{\sqrt{\sigma}},$$

$$q_2(y) = -|k^{(2)}|^2 c^2(y) - \frac{(\sigma)''}{\sqrt{\sigma}}.$$

From these equalities we can easily find $c(y)$ and $(\sigma)''/\sqrt{\sigma}$. Then solving the differential equation for $\sqrt{\sigma}$, we find $\sigma(y)$.

2.1.4. Solution of the inverse problem of acoustics by the Radon transformation

We begin with the definition of the Radon transformation. Let $g(x)$ be a function defined for all $x \in \mathbb{R}^n$, and let $g(x)$ decrease sufficiently fast at infinity. Suppose P to be an arbitrary $(n-1)$-hyperplane in $\mathbb{R}^n$. We consider $\int_P g(x)\,\mathrm{d}S$, where the integral is taken over the hyperplane P. This integral is the function of the argument P. We denote this integral by $\breve{g}(P)$ and call it the Radon transform of the function g. Note that an arbitrary hyperplane can be described by the equation

$$(x,\xi) = p, \qquad \xi = \{\xi_1,\dots,\xi_n\}, \quad \xi \neq 0,$$
$$(x,\xi) = x_1\xi_1 + x_2\xi_2 + \ldots + x_n\xi_n, \qquad p \quad \text{is a number.}$$

Therefore, $\breve{g}(P)$ can be written as $\breve{g}(\xi,p)$. We shall use an alternative definition of the Radon transformation which is not too different from the first one. Suppose that in the initial space $\mathbb{R}^{n+1}$ we have a selected direction. Then we denote the coordinate of a point along this direction by t and the other coordinates by $x = \{x_1,\dots,x_n\}$. The equation of each n-dimensional plane not parallel to the t-axes can be written as $t = p + (x,\xi)$. Here p is a fixed number equal to the coordinate of the point of intersection of the plane with the t-axis; ξ is a fixed n-dimensional vector which determines the direction of the plane. In particular, the length of the vector ξ is equal to $|\xi| = \tan\alpha$ where α is the angle between the normal vector to the plane and the t-axis. Suppose $g(x,t)$ to be a function decreasing sufficiently fast for $x \to \infty$. The function

$$\breve{g}(\xi,p) = \int g(x,p+(x,\xi))\,\mathrm{d}x \tag{2.1.11}$$

is called the Radon transform of the function g. Here the integral is taken over the whole space $\mathbb{R}^n$. Evidently, the second definition of the Radon transformation is close to the first one. The difference is in a) the way of parametrization of planes; b) the multiplier dependent on the parameter ξ. What is more, the second definition is not defined for the set of planes parallel to the t-axis.

There are special properties of the Radon transformation we should be aware of.

1. Evidently,

$$\begin{aligned}\frac{\partial}{\partial p}\breve{g}(\xi,p) &= \frac{\partial}{\partial p}\int g(x,p+(x,\xi))\,\mathrm{d}x\\ &= \int \Big(\frac{\partial}{\partial t}g(x,t)\Big)\Big|_{t=p+(x,\xi)}\,\mathrm{d}x = \Big(\frac{\breve{\partial g}}{\partial t}\Big)(\xi,p).\end{aligned}$$

2. Let us calculate

$$\Big(\frac{\partial g}{\partial x_i}\Big)\check{}(\xi,p)=\int\Big(\frac{\partial g(x,t)}{\partial x_i}\Big)\Big|_{t=p+(x,\xi)}\,dx$$
$$=\int\frac{\partial}{\partial x_i}g(x,p+(x,\xi))\,dx-\int\xi_i\Big(\frac{\partial g(x,t)}{\partial t}\Big)\Big|_{t=p+(x,\xi)}\,dx.$$

The first integral is equal to zero because $g(x,p+(x,\xi))$ decreases sufficiently fast at infinity. Therefore,

$$\Big(\frac{\partial g}{\partial x_i}\Big)\check{}(\xi,p)=-\xi_i\frac{\partial}{\partial p}\int g(x,p+(x,\xi))\,dx=-\xi_i\frac{\partial}{\partial p}\check{g}(\xi,p).$$

3. Properties 1 and 2 yield

$$\Big(\frac{\partial^2 g}{\partial t^2}\Big)\check{}(\xi,p)=\frac{\partial^2}{\partial p^2}\check{g}(\xi,p),\qquad (\Delta_x g)\check{}(\xi,p)=|\xi|^2\frac{\partial^2}{\partial p^2}\check{g}(\xi,p),\tag{2.1.12}$$

where Δ_x is the Laplace operator

$$\Delta_x g=\sum_{i=1}^{n}\frac{\partial^2}{\partial x_i^2}g.$$

4. The possibility to define $\check{g}(\xi,p)$ correctly and to prove formulas (2.1.12) is based on the assumption that the function $g(x,t)$ and its derivative decrease sufficiently fast at infinity. However, the behavior of $g(x,t)$ may be different at infinity depending on the direction. So, the integral (2.1.11) and formulas (2.1.12) make sense not for all ξ and p.[2]

Now we consider problem (2.1.1)–(2.1.4). Let $\check{u}(\xi,p,z)$ be the Radon transform of the wave field $u(x,z,t)$ regarded as a function of x,t for each fixed $z>0$. We suppose that the speed of wave propagation is bounded, $c(x)=\sqrt{b/a}\le C$; C is a constant. In this case $u\equiv 0$ for $|x|>Ct$. Hence, the Radon transform

$$\check{u}(\xi,p,z)=\int u(x,z,p+(x,\xi))\,dx$$

is defined for $|\xi|<1/C$. From formulas (2.1.12) it follows that $\check{u}(\xi,p,z)$ satisfies the differential equation

$$(a(z)-|\xi|^2 b(z))\check{u}_{pp}=\frac{\partial}{\partial z}\Big(b(z)\frac{\partial\check{u}}{\partial z}\Big),\tag{2.1.13}$$

[2] Note that the definition of the Radon transformation can be naturally extended so as to have meaning for all ξ and p. However, that definition involves more detailed application of the theory of generalized functions. We shall not use that definition here.

and the conditions

$$\breve{u}\big|_{p<0} \equiv 0, \tag{2.1.14}$$

$$b\breve{u}_z\big|_{z=0} = \delta(p), \tag{2.1.15}$$

$$\breve{u}\big|_{z=0} = \breve{f}(\xi,p), \tag{2.1.16}$$

where $\breve{f}(\xi,p)$ is the Radon transform of the function $f(x,t)$. For each fixed $|\xi|$, the problem (2.1.13)–(2.1.16) is a special case of the problem (1.1.1)–(1.1.4). Solving it, we find the function

$$\sigma_\xi(y_\xi) = \sqrt{ab - |\xi|^2 b^2}$$

as a function of the variable

$$y_\xi = \int_0^t \sqrt{a/b - |\xi|^2}\,\mathrm{d}z.$$

Assume that we have solved the inverse problem (2.1.13)–(2.1.16) for two distinct values $|\xi^{(1)}|$ and $|\xi^{(2)}|$ and have found the functions $\sigma_1(y_1)$ and $\sigma_2(y_2)$. Now we consider the question "Is it possible to recover the functions $a(z)$ and $b(z)$ from $\sigma_1(y_1)$ and $\sigma_2(y_2)$?" The answer is positive. Note that

$$\frac{\mathrm{d}y_i}{\mathrm{d}z} = \sqrt{\frac{a}{b} - |\xi^{(i)}|^2} = \frac{1}{b}\sqrt{ab - |\xi^{(i)}|^2 b^2} = \frac{\sigma_i}{b},$$

or

$$\frac{\mathrm{d}y_i}{\sigma_i(y_i)} = \frac{\mathrm{d}z}{b}. \tag{2.1.17}$$

Since the right-hand side of equation (2.1.17) is independent of i, we obtain the differential equation linking y_1 and y_2:

$$\frac{\mathrm{d}y_1}{\sigma_1(y_1)} = \frac{\mathrm{d}y_2}{\sigma_2(y_2)}.$$

This implies

$$\int_0^{y_1} \frac{\mathrm{d}y_1}{\sigma_1(y_1)} = \int_0^{y_2} \frac{\mathrm{d}y_2}{\sigma_2(y_2)}.$$

Thereby we obtain, say, $y_2(y_1)$ and hence σ_2 as a function of y_1: $\sigma_2(y_2(y_1))$. For brevity, the last function will be denoted by $\sigma_2(y_1)$. Knowing $\sigma_1^2(y_1) = ab - |\xi^{(1)}|^2 b^2$ and $\sigma_2^2(y_1) = ab - |\xi^{(2)}|^2 b^2$, we can find either of the functions $a(y_1)$ and $b(y_1)$:

$$b(y_1) = \sqrt{\frac{\sigma_1^2(y_1) - \sigma_2^2(y_1)}{|\xi^{(2)}|^2 - |\xi^{(1)}|^2}},$$

$$a(y_1) = \frac{\sigma_1^2(y_1)|\xi^{(2)}|^2 - \sigma_2^2(y_1)|\xi^{(1)}|^2}{\sqrt{(|\xi^{(2)}|^2 - |\xi^{(1)}|^2)(\sigma_1^2(y_1) - \sigma_2^2(y_1))}}.$$

These functions can also be expressed in terms of z using the relation

$$z = \int_0^y \sqrt{\frac{b}{a - |\xi^{(1)}|^2 b}}\, dy.$$

So, in order to find $a(z)$ and $b(z)$ uniquely, it suffices to know the values of $\breve{f}(\xi, p)$ for $\xi^{(1)}$ and $\xi^{(2)}$ such that $|\xi^{(1)}| \neq |\xi^{(2)}|$. Note that because of $f(x,t) = f(|x|,t)$, the Radon transform for $\xi \neq 0$ and $n = 2$ can be represented as follows:

$$\breve{f}(\xi, p) = 2\int_0^\infty r\, dr \int_{-\infty}^\infty dt f(r,t)(r^2|\xi|^2 - (t-p)^2)_+^{-1/2},$$

where

$$(s)_+^{-1/2} = \begin{cases} s^{-1/2}, & s > 0, \\ 0, & s < 0. \end{cases}$$

2.1.5. The inverse problem of scattering a plane wave

Suppose that equation (2.1.1) holds in the whole space of the variables x, z: $x \in \mathbb{R}^2$, $-\infty < z < \infty$. We suppose also that the medium is homogeneous for $z < 0$, $a(z)|_{z<0} = a_0$, $b(z)|_{z<0} = b_0$ where a_0 and b_0 are positive constants. The wave speed $c(z) = \sqrt{b/a}$ is assumed to be bounded. Let the wave field be generated by a plane wave incident from the homogeneous half-space $z < 0$:

$$u_{\text{inc}} \sim \delta(t - (x,\xi) - z). \tag{2.1.18}$$

Here $\xi = (\xi_1, \xi_2)$ ($|\xi| < 1/c_0 = \sqrt{a_0/b_0}$) is the vector determining the direction of the incident wave. Further we shall consider the case when ξ satisfies the more rigid requirement

$$|\xi| < (\sup_z c(z))^{-1}. \tag{2.1.19}$$

Violation of condition (2.1.19) means that the total interior reflection occurs at a depth of the medium. In this case, there is no solution enjoying the above properties (see [13]).

Denote $t - (x,\xi) = p$. We can show that the wave field generated by the incident wave (2.1.18) is, for a fixed ξ, a function dependent only on the arguments p and z: $u = v(p, z, \xi)$. It is easy to verify that v satisfies the equation

$$(a - |\xi|^2 b)v_{pp} = \frac{\partial}{\partial z}(bv_z),$$

which coincides with (2.1.13). This equation, for $z < 0$, has the form

$$(a_0 - |\xi|^2 b_0)v_{pp} = b_0 v_{zz}.$$

Its general solution is the sum of the incident and reflected waves:

$$\begin{aligned} v|_{z<0} &= f_{\text{inc}}\Big(p - z\sqrt{\frac{a_0}{b_0} - |\xi|^2}\Big) + f_{\text{refl}}\Big(p + z\sqrt{\frac{a_0}{b_0} - |\xi|^2}\Big) \\ &= \delta\Big(p - z\sqrt{\frac{a_0}{b_0} - |\xi|^2}\Big) + f_{\text{refl}}\Big(p + z\sqrt{\frac{a_0}{b_0} - |\xi|^2}\Big). \end{aligned}$$

The medium is assumed to be initially at rest. This means that

$$v|_{p<0} = \delta(p - z\sqrt{a_0/b_0 - |\xi|^2})$$

(for negative "time" p there exists only the wave incident from $-\infty$). These data, with known $a(z)$ and $b(z)$ and unknown f_{refl}, are sufficient for the unique determination of the wave field. Solving the direct problem, we can find $f_{\text{refl}} \stackrel{\text{def}}{=} f$. Evidently, f depends on the parameter ξ:

$$f = f(p + z\sqrt{a_0/b_0 - |\xi|^2}, \xi).$$

When setting the inverse problem, we assume that $a(z)$ and $b(z)$ are unknown, and the reflected wave f is known. In this case the one-dimensional inverse problem arisen for a fixed ξ is a special case of the problem considered in Subsection 1.6.3. Solving this problem for two different values $\xi = \xi^{(1)}$ and $\xi = \xi^{(2)}$ ($|\xi^{(1)}| \neq |\xi^{(2)}|$), we find $\sigma_1(y_1)$ and $\sigma_2(y_2)$ (see notations in Subsection 2.1.4). Then using the method presented in Subsection 2.1.4, we find $a(z)$ and $b(z)$.

2.1.6. The inverse problem of wave propagation in wave guides

Let equation (2.1.1) be satisfied in the domain $z > 0$, $x \in \Omega$, where Ω is a two-dimensional bounded domain, $\Omega \subset \mathbb{R}^2$; the boundary of Ω will be denoted by Γ. The solution of equation (2.1.1) satisfies the boundary condition

$$\alpha u + \beta \frac{\partial u}{\partial n}\Big|_{x\in\Gamma} = 0,$$

where α and β are known real-valued functions of points on the boundary, $\alpha \geq 0$, $\beta \geq 0$, $\alpha + \beta > 0$. It is assumed that

$$u\big|_{t<0} \equiv 0.$$

Suppose also that we know

$$u\big|_{z=0} = f(x,t),$$

$$u_z\big|_{z=0} = g(x,t) = g_0(x)\delta(t) + g_1(x,t).$$

Let $\{\Psi_1(x), \Psi_2(x), \ldots\}$, $\{\lambda_1, \lambda_2, \ldots\}$ be eigenfunctions and eigenvalues of the problem

$$-\Delta_x \Psi_k(x) = \lambda_k \Psi_k(x),$$

$$\alpha \Psi_k(x) + \beta \frac{\partial}{\partial n} \Psi_k(x)\Big|_{x\in\Gamma} = 0.$$

Consider

$$v_k(z,t) = \int_\Omega u(x,z,t)\Psi_k(x)\,\mathrm{d}x.$$

It is easy to verify that $v_k(z,t)$ satisfies the equation

$$a\,\frac{\partial^2 v_k}{\partial t^2} = \frac{\partial}{\partial z}\Big(b\,\frac{\partial v_k}{\partial z}\Big) - \lambda_k b v_k$$

and the conditions

$$v_k\big|_{t<0} = 0,$$

$$v_k\big|_{z=0} = F_k(t) = \int_\Omega f_k(x,t)\Psi_k(x)\,\mathrm{d}x,$$

$$v_{k\,z}\big|_{z=0} = G_k(t) = \delta(t)\int_\Omega g_0(x)\Psi_k(x)\,\mathrm{d}x + \int_\Omega g_1(x,t)\Psi_k(x)\,\mathrm{d}x.$$

Evidently (just similar to the reasoning in Subsection 2.1.3), to determine uniquely $a(z)$ and $b(z)$, it suffices to know the functions $F_k(t)$ and $G_k(t)$ for $k = 1$ and $k = 2$ provided a) $\lambda_1 \neq \lambda_2$, b) $\int_\Omega g_0(x)\Psi_k(x)\,\mathrm{d}x \neq 0$ for $k = 1$ and $k = 2$. We do not suppose here that eigenvalues are specially ordered. So, λ_1 and λ_2 may be any pair of eigenvalues.

2.1.7. The inverse problem for a layered ball

Suppose equation (2.1.1) to hold inside a ball of radius R. In this subsection the coefficients a and b are assumed to depend only on the spherical coordinate r. Equation (2.1.1) can be written in the spherical coordinate system as

$$au_{tt} = \frac{1}{r^2}\frac{\partial}{\partial r}\Big(r^2 b\frac{\partial u}{\partial r}\Big) + \frac{1}{r^2}\, b\,\Delta_{\theta,\varphi}u.$$

Here $\Delta_{\theta,\varphi}$ is the angular part of the Laplace operator; θ,φ are the angular spherical coordinates. As usual, we assume

$$u\big|_{t<0} \equiv 0,$$

$$u\big|_{r=R} = f(\theta,\varphi,t),$$

$$bu_r\big|_{r=R} = g(\theta,\varphi,t) = g_0(\theta,\varphi)\delta(t) + g_1(\theta,\varphi,t).$$

Let $Y_{lm}(\theta,\varphi)$ be a spherical harmonic. Consider

$$u_{lm}(r,t) = \int_0^{\pi} \mathrm{d}\theta \int_0^{2\pi} \mathrm{d}\varphi u(r,\theta,\varphi,t)Y_{lm}(\theta,\varphi)\sin\theta$$

or, briefly,

$$u_{lm}(r,t) = \iint uY_{lm}\sin\theta\,\mathrm{d}\theta\,\mathrm{d}\varphi.$$

As is easy to see, the equality

$$\Delta_{\theta,\varphi}Y_{lm}(\theta,\varphi) = -l(l+1)Y_{lm}(\theta,\varphi)$$

implies that $u_{lm}(r,t)$ satisfies the differential equation

$$a\frac{\partial^2}{\partial t^2}u_{lm} = \frac{1}{r^2}\frac{\partial}{\partial r}\Big(r^2 b\frac{\partial}{\partial r}u_{lm}\Big) - \frac{l(l+1)}{r^2}bu_{lm}.$$

We go now to a new variable z, $r = R\mathrm{e}^{-z}$ $(z > 0)$, and denote

$$Ra\mathrm{e}^{-3z} = \tilde{a}, \qquad \mathrm{e}^{-z}b = \tilde{b}R.$$

In these new variables the equation evidently takes the form

$$\tilde{a}\frac{\partial^2}{\partial t^2}u_{lm} = \frac{\partial}{\partial z}\Big(\tilde{b}\frac{\partial u_{lm}}{\partial z}\Big) - l(l+1)\tilde{b}u_{lm}, \qquad z > 0.$$

It is easy to verify that u_{lm} satisfies the initial condition

$$u_{lm}\big|_{t<0} \equiv 0$$

and the boundary conditions

$$u_{lm}|_{z=0} = f_{lm}(t),$$

$$\tilde{b}\,\frac{\partial u_{lm}}{\partial z}\bigg|_{z=0} = g_{0lm}\delta(t) + g_{1lm}(t),$$

where

$$f_{lm}(t) = \iint f Y_{lm} \sin\theta \, \mathrm{d}\theta \, \mathrm{d}\varphi, \qquad g_{0lm} = \iint g_0 Y_{lm} \sin\theta \, \mathrm{d}\theta \, \mathrm{d}\varphi,$$

$$g_{1lm}(t) = \iint g_1 Y_{lm} \sin\theta \, \mathrm{d}\theta \, \mathrm{d}\varphi.$$

If we know the functions $f_{lm}(t)$ and $g_{1lm}(t)$ and the constants $g_{0lm} \neq 0$ for two pairs of indices l_1, m_1 and l_2, m_2, with $l_1 \neq l_2$, then, using the same scheme as in the previous cases, we recover uniquely the coefficients $\tilde{a}(z)$ and $\tilde{b}(z)$ and therefore $a(r)$ and $b(r)$.

2.1.8. The method of moments. Formulation of a problem and its reduction to an integral equation

Of course, we may consider some more cases when the problem of recovering the coefficients of (2.1.1) dependent on a single coordinate is solved by the methods analogous to those considered above. However, we restrict ourselves to the above examples. Further, we shall propose another method of solution of the problem (2.1.1)–(2.1.4). After applying the Fourier transformation with respect to x and changing from z to the coordinate y, the problem takes the form

$$\tilde{u}_{tt} = \frac{1}{\sigma(y)}\frac{\partial}{\partial y}\left(\sigma(y)\frac{\partial \tilde{u}}{\partial y}\right) - |k|^2\,\frac{b}{a}\,\tilde{u}, \tag{2.1.20}$$

$$\tilde{u}|_{t<0} = 0, \qquad \sigma\tilde{u}_y|_{y=0} = \delta(t), \tag{2.1.21}$$

$$\tilde{u}|_{y=0} = \tilde{f}(k,t). \tag{2.1.22}$$

We consider small values of the parameter $|k|^2$ and replace the function $\tilde{u}$ by its expansion[3] in powers of $|k|^2$:

$$\tilde{u} = \tilde{u}_0(y,t) + |k|^2\tilde{u}_1(y,t) + \dots . \tag{2.1.23}$$

[3]Since the speed of wave propagation is finite, the function $\tilde{u}$ is the Fourier transform of a function finite for each fixed values of y and t. Therefore, the function is analytic in k. The problem has axial symmetry; therefore, u is a function of $|x|$ (for fixed y and t) and $\tilde{u}$ is a function of $|k|$. Functions analytic in k and dependent only on $|k|$ can be expanded into a series of the form (2.1.23). Evidently, the coefficients in this expansion are the

Similarly, we can expand $\tilde{f}(k,t)$:

$$\tilde{f} = \tilde{f}_0(t) + |k|^2 \tilde{f}_1(t) + \dots. \tag{2.1.24}$$

Substituting these expansions into (2.1.20)–(2.1.22) and equating the terms with the same powers of $|k|^2$, we obtain, in particular, two inverse problems. The former is as follows:

$$\tilde{u}_{0tt} = \frac{1}{\sigma(y)} \frac{\partial}{\partial y}(\sigma(y)\tilde{u}_{0y}), \tag{2.1.25}$$

$$\tilde{u}_0\big|_{t<0} = 0, \qquad \sigma \tilde{u}_{0y}\big|_{y=0} = \delta(t), \tag{2.1.26}$$

$$\tilde{u}_0\big|_{y=0} = \tilde{f}_0(k,t). \tag{2.1.27}$$

This problem, evidently, coincides with the inverse problem from Subsection 2.1.3 for $k = 0$. Suppose it has been solved and the function $\sigma(y)$ has been obtained. In this case, the function $\tilde{u}_0$ can also be found (without additional calculations if we use the method of nonlinear Volterra equations, see Section 1.3; or by formula (1.6.6) if the method of Gel'fand–Levitan equations is applied).

We consider now the latter inverse problem resulted from equating the terms containing $|k|^2$:

$$\tilde{u}_{1tt} = \frac{1}{\sigma(y)} \frac{\partial}{\partial y}(\sigma(y)\tilde{u}_{1y}) - \frac{b}{a}\tilde{u}_0, \tag{2.1.28}$$

$$\tilde{u}_1\big|_{t<0} = 0, \qquad \sigma \tilde{u}_{1y}\big|_{y=0} = 0, \tag{2.1.29}$$

$$\tilde{u}_1\big|_{y=0} = \tilde{f}_1(t). \tag{2.1.30}$$

In this problem the functions $\tilde{u}_1$ and $q = b/a$ are unknown. Unlike all the above problems, the inverse problem (2.1.28)–(2.1.30) is linear. We shall reduce it to a linear Volterra integral equation. Introduce the Green function as the solution of the equation

$$G_{tt} - \frac{1}{\sigma(y)} \frac{\partial}{\partial y}\Big(\sigma(y)\frac{\partial}{\partial y}G(y)\Big) = \delta(y - y', t), \tag{2.1.31}$$

---

moments of the function u:

$$\tilde{u}_0 = \int e^{i(k,x)} u(x,z,t)\, dx\Big|_{k=0} = \int u(x,z,t)\, dx,$$

$$\tilde{u}_1 = \frac{1}{2}\sum_{j=1}^{2} \frac{\partial^2}{\partial k_j^2} \int e^{i(k,x)} u(x,z,t)\, dx\Big|_{k=0} = -\frac{1}{2}\int |x|^2 u(x,z,t)\, dx, \dots.$$

satisfying the conditions

$$G|_{t<0} = 0, \qquad \sigma G_y|_{y=0} = 0. \tag{2.1.32}$$

If G is known, we can construct easily a solution of the equation

$$\tilde{v}_{1tt} - \frac{1}{\sigma(y)}\Big(\frac{\partial}{\partial y}\sigma(y)\frac{\partial \tilde{v}}{\partial y}\Big) = h(y,t),$$

with an arbitrary right-hand side $h(y,t)$ ($h \equiv 0$ for $t < 0$), satisfying conditions (2.1.32). This solution is as follows:

$$v(y,t) = \int_0^\infty \mathrm{d}y'\, G(y,y',t) * h(y',t).$$

Here the star, as before, denotes the convolution with respect to time:

$$G(y,y',t) * h(y',t) = \int_0^\infty G(y,y',t-t')h(y',t')\,\mathrm{d}t'.$$

In particular, substituting the function $(-q\tilde{u}_0)$ for h, we obtain the relation

$$\tilde{u}_1(y,t) = -\int_0^\infty \mathrm{d}y'\, G(y,y',t) * \tilde{u}_0(y',t)q(y'). \tag{2.1.33}$$

Note that both the integrals, with respect to y' and with respect to t' in the convolution, are taken between finite limits. This follows from the fact that velocity of wave propagation is finite. We set $y = 0$ in (2.1.33) to obtain the integral equation

$$\tilde{f}_1(0,t) = \int_0^\infty \mathrm{d}y'\, G(0,y',t) * \tilde{u}_0(y',t)q(y'). \tag{2.1.34}$$

2.1.9. Construction of the Green function

It is remarkable that the Green function $G(y,y',t)$ can be constructed explicitly if the functions $\tilde{u}_0(y,t)$ and $V(y,t)$ are known. Recall that $V(y,t)$ is derived by formula (1.6.5) from the solution $v(y,t)$ of the Gel'fand – Levitan equation with $\tilde{f}_0(t)$ instead of f. Note that both $\tilde{u}$ and V satisfy equation (2.1.25).

The construction of the Green function is based on the following considerations. We regard the Green function as a generalized function of t;

y and y' are parameters. Then for $y > y'$ and $y < y'$, the Green function satisfies the homogeneous equation

$$G_{tt} = \frac{1}{\sigma(y)} \frac{\partial}{\partial y}\Big(\sigma(y)\frac{\partial}{\partial y} G(y,y',t)\Big). \tag{2.1.35}$$

For $y = y'$, the Green function is continuous,

$$\lim_{y\to y'+0} G(y,y',t) = \lim_{y\to y'-0} G(y,y',t). \tag{2.1.36}$$

and meets the condition

$$\lim_{y\to y'+0} \frac{\partial}{\partial y} G(y,y',t) - \lim_{y\to y'-0} \frac{\partial}{\partial y} G(y,y',t) = -\delta(t). \tag{2.1.37}$$

In both the equalities the limit is understood in the sense of the theory of generalized functions.

For $y > y'$, the Green function has to model the wave running from the source at y' to the right. A solution of equation (2.1.35) of such type is known. It is the function $\tilde{u}_0(y,t)$. Evidently, any function of the form $C_1(t) * \tilde{u}_0(y,t)$ is also a solution of this type. For $y < y'$, the Green function, as a function of y, must satisfy condition (2.1.32) for $y = 0$. A solution of such type is also known, namely, the function $V(y,t)$ (or its convolution with an arbitrary function of t). These considerations show that it is reasonable to seek for the Green function in the form

$$G(y,y',t) = \begin{cases} C_1(y',t) * \tilde{u}_0(y,t), & \text{if} \quad y > y', \\ C_2(y',t) * V(y,t), & \text{if} \quad y < y'. \end{cases} \tag{2.1.38}$$

Evidently, the functions C_1 and C_2 must depend on y', which is taken into account in formula (2.1.38). Conditions (2.1.36), (2.1.37) combined with (2.1.38) result in the system of equations for C_1 and C_2:

$$C_1(y,t) * \tilde{u}_0(y,t) - C_2(y,t) * V(y,t) = 0,$$

$$C_1(y,t) * \tilde{u}_0'(y,t) - C_2(y,t) * V'(y,t) = -\delta(t).$$

Here we denote by the prime the differentiation with respect to y. Convolving the first equation with $V'(y,t)$, the second with $V(y,t)$, and subtracting one from another, we find

$$C_1(y,t) * W(y,t) = -V(y,t),$$

where $W = \tilde{u}_0' * V - \tilde{u}_0 * V'$. Analogously, we can get

$$C_2(y,t) * W(y,t) = -\tilde{u}_0(y,t).$$

Note that the function W is an analog of the Wronskian. Let us find now W. For this purpose, we calculate $(\sigma W)'$. Recall that a convolution is differentiated with respect to a parameter following the same rule as a product, that is,

$$\frac{\mathrm{d}}{\mathrm{d}y}(\tilde{u}_0 * V) = \tilde{u}_0' * V + \tilde{u}_0 * V'.$$

The differentiation with respect to t is performed by the rule

$$\frac{\partial}{\partial t}(\tilde{u}_0 * V) = \left(\frac{\partial \tilde{u}_0}{\partial t}\right) * V = \tilde{u}_0 * \left(\frac{\partial V}{\partial t}\right).$$

These rules were actually used before. However, in the below calculation they will be exploited more intensively. So we have

$$\begin{aligned}\frac{\partial}{\partial y}(\sigma W) &= \frac{\partial}{\partial y}((\sigma \tilde{u}_0') * V) - \frac{\partial}{\partial y}(\tilde{u}' * (\sigma V')) \\ &= (\sigma \tilde{u}_0')' * V + \sigma \tilde{u}_0' * V' - \tilde{u}_0' * \sigma V' - \tilde{u}_0 * (\sigma V')' \\ &= \sigma(\tilde{u}_{0tt}' * V - \tilde{u}_0' * V_{tt}) = \sigma \frac{\partial^2}{\partial t^2}(\tilde{u}_0 * V - \tilde{u}_0 * V) = 0.\end{aligned}$$

Here we have used the fact that $\tilde{u}_0$ and V satisfy equation (2.1.25). The last equality yields that σW is independent of y. In order to calculate σW, we take into account the following condition for $y = 0$:

$$\sigma W|_{y=0} = (\sigma \tilde{u}_0' * V - \tilde{u}_0 \sigma * V')|_{y=0} = \delta(t) * \delta(t) - 0 = \delta(t).$$

As a result, equations for C_1 and C_2 get trivial form

$$C_1(y,t) * \sigma^{-1}(y)\delta(t) = -V(y,t), \qquad C_2(y,t) * \sigma^{-1}(y)\delta(t) = -\tilde{u}_0(y,t),$$

whence

$$C_1(y,t) = -\sigma(y)V(y,t), \qquad C_2(y,t) = -\sigma(y)\tilde{u}_0(y,t).$$

The final expression for the Green function is as follows:

$$G(y,y',t) = \begin{cases} -\sigma(y')V(y',t) * \tilde{u}_0(y,t), & \text{if} \quad y > y', \\ -\sigma(y')\tilde{u}_0(y',t) * V(y,t), & \text{if} \quad y < y'. \end{cases}$$

We can show that from the explicit form of the Green function it follows that

$$G(y,y',t)\equiv 0 \quad \text{for} \quad t<|y-y'|.$$

From the physical point of view, it is evident. Really, the wave field of an instant point source is equal to zero until the wave generated by the source comes.

2.1.10. Investigation of integral equation (2.1.34)

Let us rewrite equation (2.1.34) taking into account that the first argument of the Green function is equal to zero and using the expression for the Green function which holds for $y<y'$. As a result, we obtain

$$\tilde{f}_1(t)=-\int_0^\infty \sigma(y')\tilde{u}_0(y',t)*V(0,t)*\tilde{u}_0(y',t)q(y')\,\mathrm{d}y'.$$

This equation is conveniently written as follows:

$$\tilde{f}_1(2t)=\int_0^\infty T(y,t)q(y)\,\mathrm{d}y, \tag{2.1.39}$$

where

$$T(y,t)=-\sigma(y)\tilde{u}_0(y,t_1)*V(0,t_1)*\tilde{u}_0(y,t_1)|_{t_1=2t}\,.$$

Recalling that $V(0,t)=\delta(t)$, we find that

$$T(y,t)=\sigma(y)\tilde{u}_0(y,t_1)*\tilde{u}_0(y,t_1)|_{t_1=2t}\,. \tag{2.1.40}$$

Note that $\tilde{u}_0(y,t)\equiv 0$ for $t<y$. Therefore,

$$\tilde{u}_0(y,t_1)*\tilde{u}_0(y,t_1)|_{t_1=2t}=\int_y^{2t-y}\tilde{u}_0(y,2t-t')\tilde{u}_0(y,t')\,\mathrm{d}t'$$

if $y<t$; otherwise this convolution is equal to zero. Taking into account the behavior of the function $\tilde{u}_0(y,t)$ for t close to y:

$$\tilde{u}_0(y,t)=-\frac{\varepsilon(t-y)}{\sqrt{\sigma(y)\sigma(0)}}+\dots,$$

we obtain the singularity of the kernel T: for $t>y$,

$$T(y,t)\sim\frac{\sigma(y)}{\sigma(y)\sigma(0)}\int\varepsilon(2t-t'-y)\varepsilon(t'-y)\,\mathrm{d}t'=\frac{1}{\sigma(0)}\int_y^{2t-y}\mathrm{d}t'.$$

Hence,

$$T(y,t) = \frac{2}{\sigma(0)}(t-y)_+ + \dots .$$

Double differentiation of equation (2.1.39) with respect to t yields

$$\ddot{f}_1(2t) = \frac{2}{\sigma(0)} q(t) + \int_0^t \ddot{T}(t,y)q(y)\,dy. \qquad (2.1.41)$$

The dot denotes differentiation with respect to t. We denote by $\ddot{T}$ the function $T(t,y)$ differentiated with respect to the first argument for $t > y$ (this means that $\ddot{T}$ has no δ-term resulted from the double differentiation of T; when integrated this term causes the first term of (2.1.41)). Equation (2.1.41) is the Volterra integral equation of the second kind. From this it follows that there exists a unique solution which can be constructed using the method of successive approximations. It occurs that the result can be improved. Introduce the operator

$$\hat{T} = \int_0^\infty T(y,t)(\cdot)\,dy.$$

It is remarkable that $\hat{T}^{-1}$ can be obtained explicitly.

2.1.11. The inversion formula for the operator $\hat{T}$

Introduce the operator

$$\hat{S} = -4\int_0^{y+0} \frac{\partial^2}{\partial s\,\partial y} S(y,s)(\cdot)\,ds,$$

where

$$S(y,s) = \sigma V(y,t) * V(y,t)|_{t=2s} \cdot$$

Here V is the function defined in Subsection 1.6.1. Recall that V is determined from the solution of the Gel'fand–Levitan equation.

Let us show now that $\hat{S} = \hat{T}^{-1}$. We shall give here only the main points of the proof. For simplicity, we shall assume that $\sigma(y)$ is three times continuously differentiable. It is easy to verify that T and S solve the differential equation

$$T_{tty} = \mathcal{L}T, \qquad S_{tty} = \mathcal{L}S,$$

where $\mathcal{L}$ is the following differential operator:

$$\mathcal{L}\varphi = \varphi_{yyy} - 4Q\varphi_y - 2Q_y\varphi, \qquad Q = \frac{(\sqrt{\sigma})_{yy}}{\sqrt{\sigma}}.$$

These equations follow from the fact that the functions S, T are the products of $\sigma(y)$ by the convolutions of functions ($\tilde{u}_0$ or V) satisfying equation (2.1.25). The formula of integration by parts yields the following relation for each pair of functions $\varphi(y)$ and $\psi(y)$:

$$\int_0^a \mathcal{L}\varphi(y)\psi(y)\,dy = -\int_0^a \varphi(y)\mathcal{L}\psi(y)\,dy + (\varphi_{yy}\psi - \varphi_y\psi_y + \varphi\psi_{yy} - 4q\varphi\psi)\Big|_0^a.$$

Consider the function

$$\Phi(s,t) = \int_0^\infty S_y(y,s)T(y,t)\,dy = -\int_0^\infty S(y,s)T_y(y,t)\,dy - S(0,s)T(0,t).$$

Using the equations for S and T and the formula of integration by parts for the operator $\mathcal{L}$, we can check that the function $\Phi(s,t)$ satisfies the partial differential equation

$$\Phi_{tt} - \Phi_{ss} = -\frac{\sigma(0)}{4}\Big(\frac{\partial^2}{\partial t^2} - \frac{\partial^2}{\partial s^2}\Big)T(0,t)\delta(s) + \frac{1}{2}\delta(s)\delta(t).$$

Note that, because of $\tilde{u}_0|_{t<0} = 0$, we have $T|_{t<0} = 0$ and $\Phi|_{t<0} = 0$. Hence,

$$\Phi = -\frac{\sigma(0)}{4}T(0,t)\delta(s) + \frac{1}{4}\varepsilon(t - |s|).$$

Therefore, for $s > 0$,

$$\Phi = \varepsilon(t-s); \qquad -\frac{\partial\Phi}{\partial s} = \frac{1}{4}\delta(t-s).$$

Now we calculate the product of the operators $\hat{T}$ and $\hat{S}$:

$$\begin{aligned}\hat{T}\hat{S}\psi(t) &= -4\int_0^t T(y,t)\,dy \int_0^{y+0} \frac{\partial^2}{\partial s\,\partial y}S(y,s)\psi(s)\,ds\\ &= -4\int_0^\infty \Big(\int_0^\infty T(y,t)\frac{\partial^2}{\partial s\,\partial y}S(y,s)\,dy\Big)\psi(s)\,ds\\ &= -4\int_0^\infty \frac{\partial}{\partial s}\Phi(t,s)\psi(s)\,ds = \int_0^\infty \delta(t-s)\psi(s)\,ds = \psi(t).\end{aligned}$$

Hence, $\hat{S} = \hat{T}^{-1}$, which was to be proved.

2.1.12. Once more about the scattering problem

It was shown in Subsection 2.1.5 that the inverse scattering problem for a plane wave incident from a homogeneous half-space on an inhomogeneous layered one can be reduced to the one-dimensional problem for the equation

$$(a - |\xi|^2 b)v_{pp} = \frac{\partial}{\partial z}(bv_z), \qquad (z,p) \in \mathbb{R}^2,$$

subject to the condition

$$v|_{p<0} = \delta\Big(\frac{p}{\sqrt{1-\xi^2}} - z\Big).$$

We also assume[4] that

$$a(z)|_{z<0} = a_0 = 1; \qquad b(z)|_{z<0} = b_0 = 1,$$

$a(z) > 0$, $b(z) > 0$ are continuous, and $a(z) - |\xi|^2 b(z) > 0$ for all z.

It is convenient to introduce new coordinates

$$x = \int_0^z \frac{dz}{b(z)}, \qquad t = \frac{p}{\sqrt{1-|\xi|^2}}.$$

Upon changing to the new variables, the equation and the condition for $t < 0$ become as follows:

$$(A + \lambda B)v_{tt} = v_{xx}, \qquad (x,t) \in \mathbb{R}^2, \tag{2.1.42}$$

$$v|_{t<0} = \delta(t - x), \tag{2.1.43}$$

where

$$A = ab, \qquad B = b(a-b), \qquad \lambda = \frac{|\xi|^2}{1-|\xi|^2}.$$

Evidently, $A + \lambda B = 1$ for $x < 0$. Hence, taking into account (2.1.43), we obtain

$$v|_{x<0} = \delta(t-x) + f(t+x). \tag{2.1.44}$$

Here the function f, which describes the reflected wave, depends on the parameter λ characterizing the angle of incidence of the plane wave. In Subsection 2.1.5 we have considered the problem (in terms of equation (2.1.42)) of determination of the coefficients A and B if the function f is given for two values of λ. The inverse problem can be set in another form: recover

[4]The fact that $a_0 = b_0 = 1$ does not imply a loss of generality.

the coefficients A and B, given the values of $f(t)$ and $g(t) = \partial f(t)/\partial\lambda$ for a fixed $\lambda = \lambda_0$. In what follows we shall assume, for simplicity, $\lambda_0 = 0$ (which corresponds to the normal incidence of the wave). The last restriction is evidently inessential.

So, for $\lambda = 0$, we arrive at the following problem for the function v:

$$Av_{tt} = v_{xx}, \qquad v|_{t<0} = \delta(t-x), \quad v|_{x<0} = \delta(t-x)+f(t+x), \qquad (x,t) \in \mathbb{R}^2. \tag{2.1.45}$$

For $v_1 = \partial v/\partial\lambda$, we get

$$Av_{1xx} = v_{1xx} - Bv_{tt}, \qquad v_1|_{t<0} = 0, \quad v_1|_{x<0} = g(t+x). \tag{2.1.46}$$

To obtain A, we solve the inverse problem (2.1.45). Upon changing from x to the coordinate $y = \int_0^x \sqrt{A}\,\mathrm{d}x$, the problem (2.1.45) takes the form

$$U_{tt} = \frac{1}{\sigma}\frac{\partial}{\partial y}(\sigma U_y), \qquad U|_{t<0} = \delta(t-y), \qquad U|_{y<0} = \delta(t-y) + f(t+y). \tag{2.1.47}$$

Here $U = U(y,t) = v(x(y),t)$, $\sigma = \sqrt{A}$. Evidently, the inverse problem (2.1.47) is a special case of the problem considered in Subsection 1.6.3.

Suppose that the problem (2.1.47) has been solved, and the function $\sigma(y)$ has been determined. Moreover, when solving (2.1.47), we have also found the solution $w(y,t)$ of the Gel'fand–Levitan equation (1.6.15) and constructed the functions

$$W(y,t) = \frac{1}{\sqrt{\sigma(y)}}(\delta(t-y) + w(y,t)) \tag{2.1.48}$$

and $U(y,t)$ (see (1.6.14)). We pass on to the problem (2.1.46). Rewriting this problem in terms of y results in

$$v_{1tt} - \frac{1}{\sigma}\frac{\partial}{\partial y}(\sigma v_{1y}) = -BU_{tt}, \qquad v_1|_{t<0} = 0, \qquad v_1|_{y<0} = g(t+y). \tag{2.1.49}$$

We invert the operator in the left-hand side of equation (2.1.49) to obtain

$$v_1(y,t) = -\int G(y,\eta,t-\tau)B_1(\eta)U_{\tau\tau}(\tau,\eta)\,\mathrm{d}\eta\,\mathrm{d}\tau, \tag{2.1.50}$$

where $G(y,\eta,t)$ is a solution of the equation

$$G_{tt} - \frac{1}{\sigma}\frac{\partial}{\partial y}(\sigma G_y) = \delta(t)\delta(y-\eta), \qquad G|_{t<0} = 0.$$

(The function G is determined from the function σ obtained before.[5]) Setting $y = 0$ in (2.1.50), we obtain the linear equation for B:

$$g(t) = -\int G(0, \eta, t - \tau) B(\eta) U_{\tau\tau}(\eta, \tau) \, d\eta \, d\tau.$$

Integrating this equation with respect to t and replacing t by $2t$, we get

$$\begin{aligned}\int_0^{2t} g(\tau) \, d\tau &= -\int_0^{\infty} d\eta \Big(\int d\tau \, G(0, \eta, 2t - \tau) U_\tau(\tau, \tau) \Big) B(\eta) \\ &= -\int_0^{\infty} d\eta \Big(\int d\tau \, \dot{G}(0, \eta, 2t - \tau) U(\tau, \tau) \Big) B(\eta). \quad (2.1.51)\end{aligned}$$

Here the dot denotes differentiation with respect to the last argument. The function $G(y, \eta, t)$ can be written explicitly as

$$\dot{G}(y, \eta, t) = \frac{\sigma(\eta)}{2} \begin{cases} W^-(\eta) * U(y), & y > \eta, \\ U(\eta) * W^-(y), & y < \eta. \end{cases} \quad (2.1.52)$$

The star denotes the convolution with respect to time. Thus, equation (2.1.51) takes the form

$$\int_0^{2t} g(\tau) \, d\tau = -\frac{1}{2} (\hat{T} B)(t) = -\frac{1}{2} \int_0^{\infty} T(y, 2t) B(y) \, dy, \quad (2.1.53)$$

where T is defined by the formula

$$T(y, t) = \sigma(y) U(y, t) * U(y, t). \quad (2.1.54)$$

Taking into account the properties of U, we rewrite equation (2.1.53) in the form

$$\int_0^{2t} g(\tau) \, d\tau = -\frac{1}{2} B(t) - \frac{1}{2} \int_0^{t-0} T(y, 2t) B(y) \, dy, \quad (2.1.55)$$

with $T(y, 2t)$ being continuous for $y < t$.

[5]Note that these calculations are similar to those performed in Subsections 2.1.8–2.1.11. Therefore, we only outline the calculations without going into detail. By the same reason, we keep the notations accepted in Subsections 2.1.8–2.1.11 for the function playing similar roles.

Thus, equation (2.1.53) is a linear Volterra integral equation of second kind, which implies its unique solvability. What is more, we can represent the operator $\hat{T}^{-1}$ in an explicit form using the identity

$$\int_{-\infty}^{\infty} \Psi'(y,t)T(y,\tau)\,\mathrm{d}y = -\frac{\mathrm{d}}{\mathrm{d}t}\delta(t-\tau), \tag{2.1.56}$$

where

$$\Psi(y,t) = \sigma(y)W(y) * W(y);$$

the prime denotes differentiation with respect to y. Via (2.1.56), we can show that

$$B(y) = -2\int_0^{\infty} \mathrm{d}\tau\, \Psi'(y,\tau)\int_0^{\tau} (\tau - s)g(s)\,\mathrm{d}s, \tag{2.1.57}$$

which is the desired solution of the problem in question.

2.2. GENERAL SECOND-ORDER HYPERBOLIC EQUATION. PROBLEM IN A HALF-SPACE

2.2.1. Setting of a problem

Suppose that the function $u(x,z,t)$, where $x = (x_1,\dots,x_n)$, $-\infty < x_k < \infty$, $z > 0$, satisfies the equation

$$u_{tt} = a_{00}\frac{\partial^2 u}{\partial z^2} + 2\sum_{j=1}^{n} a_{0j}\frac{\partial^2 u}{\partial z\partial x_j} + \sum_{j,l=1}^{n} a_{jl}\frac{\partial^2 u}{\partial x_j\partial x_l} + b_0\frac{\partial u}{\partial z} + \sum_{j=1}^{n} b_j\frac{\partial u}{\partial x_j} + cu. \tag{2.2.1}$$

All the coefficients a_{00}, a_{0j}, a_{jl}, b_0, b_j, c are assumed to depend only on the coordinate z, with equation (2.2.1) being hyperbolic: the quadratic form $\sum_{i,j=0}^{n} a_{ij}(z)k_ik_j$ is positive for any k_i $(i = 0,1,\dots,n)$ not all 0. The function u satisfies the trivial initial condition

$$u|_{t<0} \equiv 0 \tag{2.2.2}$$

and the boundary conditions

$$u_z|_{z=0} = \delta(x,t), \tag{2.2.3}$$

$$u|_{z=0} = f(x,t). \tag{2.2.4}$$

Evidently, the problem (2.2.1)–(2.2.4) with known coefficients is overdetermined; for unique determination of $u(x,z,t)$ it suffices to know either condition (2.2.3) or (2.2.4). Therefore, we can set an inverse problem: given conditions (2.2.3) and (2.2.4), recover all the coefficients of equation (2.2.1). This problem is analogous to the problem (2.1.1)–(2.1.4), but it has more unknown coefficients, namely: the $(n+1)(n+2)/2$ coefficients of the quadratic form (the coefficients a_{jl}, $u, j = 0,\dots,n$), the $n+1$ coefficients b_0, b_j, and the coefficient c. So, we have to recover $(n+2)(n+3)/2$ functions. Nevertheless, the problem remains overdetermined: we need to find a finite number of functions of one variable, given a function of $(n+1)$ variables (or an n-dimensional continuum of functions of one variable). Therefore, it is natural to expect that the inverse problem has a unique solution. However, this turns out to be wrong. Really, the below consideration of the direct problem (2.2.1)–(2.2.3) shows that the function $f(x,t)$ is determined not from all the set of coefficients (a_{ij}, b_i, c) ($i = 0,\dots,n$; $j = 0,\dots,n$) but from a collection of $(n+1)(n+2)/2$ combinations of these coefficients. Therefore, if we vary the coefficients so as to keep the combinations we shall obtain the same function $f(x,t)$. This means that one function $f(x,t)$ may correspond to different sets of the coefficients of equation (2.2.1).

2.2.2. Transformation of the problem

We shall prove now the statement formulated above. To this end, we use the Fourier transform with respect to the variable x. Let

$$\tilde{u}(k,z,t) = \int \mathrm{e}^{\mathrm{i}(k,x)} u(x,z,t)\,\mathrm{d}x.$$

Here $k = (k_1,\dots,k_n)$, $(k,x) = \sum k_j x_j$, $\int(\cdot)\,\mathrm{d}x$ is the abbreviation of the integral $\int_{-\infty}^{+\infty}\dots\int_{-\infty}^{+\infty}(\cdot)\,\mathrm{d}x_1\,\mathrm{d}x_2\dots\,\mathrm{d}x_n$. The function $\tilde{u}(k,z,t)$ satisfies the equation

$$\tilde{u}_{tt} = a_{00}\tilde{u}_{zz} - 2\mathrm{i}\sum a_{0j}k_j\tilde{u}_z - \sum_{j,l} a_{jl}k_jk_l\tilde{u} + b_0\tilde{u}_z - \mathrm{i}\sum b_jk_j\tilde{u} + c\tilde{u}$$

and the conditions

$$\tilde{u}|_{t<0} = 0, \qquad \tilde{u}_z|_{z=0} = \delta(t).$$

We introduce the coordinate $y = \int_0^z \mathrm{d}z/\sqrt{a_{00}}$ and change from z to y. Then the equation becomes as follows:

$$\tilde{u}_{tt} = \sqrt{a_{00}}\frac{\partial}{\partial y}\Big(\frac{1}{\sqrt{a_{00}}}\frac{\partial \tilde{u}}{\partial y}\Big) - \frac{2\mathrm{i}}{\sqrt{a_{00}}}\sum a_{0j}k_j\tilde{u}_y$$

$$+\frac{b_0}{\sqrt{a_{00}}}\tilde{u}_y+\Big(c-\mathrm{i}\sum b_jk_j-\sum_{j,l}a_{jl}k_jk_l\Big)\tilde{u}.$$

Also, we introduce a new function $\tilde{v}=s(y)\tilde{u}$, with the multiplier $s(y)$ so chosen that after the change the equation has no terms with the first derivative of $\tilde{v}$. For this purpose, we should set

$$s=\Big(\frac{a_{00}}{a_{00}(0)}\Big)^{-1/4}\exp\Big(-\int_0^y\Big(\mathrm{i}\sum a_{0j}k_j-\frac{1}{2}b_0\Big)a_{00}^{-1/2}\,\mathrm{d}y\Big).$$

Due to the choice of the constant of integration, we have $s|_{y=0}=1$. Straightforward computations, while rather cumbersome, result in the differential equation for the function $\tilde{v}$:

$$\tilde{v}_{tt}=\tilde{v}_{yy}-q(y,k)\tilde{v}, \tag{2.2.5}$$

where

$$q(y,k)=q_0(y)-\mathrm{i}\sum_{j=1}^{n}q_jk_j-\sum_{j,l=1}^{n}q_{jl}k_jk_l,$$

$$q_0=c-a_{00}^{1/4}(a_{00}^{-1/4})''_{yy}-\frac{\sqrt{a_{00}}}{2}\Big(\frac{b_0}{a_{00}}\Big)'_y-\frac{b_0^2}{4a_{00}}, \tag{2.2.6}$$

$$q_j=b_j-\frac{b_0}{a_{00}}a_{0j}-\sqrt{a_{00}}\Big(\frac{a_{0j}}{a_{00}}\Big)', \tag{2.2.7}$$

$$q_{jl}=a_{jl}-\frac{a_{0j}a_{0l}}{a_{00}}. \tag{2.2.8}$$

The function $\tilde{v}$ satisfies the conditions

$$\tilde{v}|_{t<0}=0,$$

$$\Big(\tilde{v}'+\Big(\lambda_0-\mathrm{i}\sum_{j=1}^{n}\lambda_jk_j\Big)\tilde{v}\Big)\Big|_{y=0}=\mu_0\delta(t), \tag{2.2.9}$$

$$\lambda_0=\frac{1}{4}\frac{a'_{00}}{a_{00}}-\frac{b_0}{2\sqrt{a_{00}}}\Big|_{y=0},\quad \lambda_j=-\frac{a_{0j}}{2\sqrt{a_{00}}}\Big|_{y=0},\quad \mu=\sqrt{a_{00}(0)}. \tag{2.2.10}$$

Evidently, the function $\tilde{v}$ is determined uniquely by the collection of the functions $q_0(y)$, $q_j(y)$, $q_{jl}(y)$ and the constants λ_0,λ_j $(j=1,\dots,n)$, μ_0. Therefore, the same collection determines uniquely $\tilde{f}(k,t)=\tilde{v}(k,y,t)|_{y=0}$ and $f(x,t)$.

Thus, as far as the inverse problem is concerned, it is reasonable to find only those quantities which $f(x,t)$ depends on, namely: the $(n+1)(n+2)/2$

functions q_0, q_j, q_{jl} and the $(n+2)$ constants $\lambda_0, \lambda_j, \mu_0$. Using the methods of Chapter 1 (we especially recommend the method of moments to be applied to this case) it is not difficult to show that all the quantities $q_0, q_j, q_{jl}, \lambda_0, \lambda_j, \mu_0$ can be uniquely determined from $\tilde{f}(k,t)$. Evidently, knowledge of these quantities is not sufficient to recover uniquely all $(n+2)(n+3)/2$ coefficients of the initial equation.

2.3. THE SCATTERING PROBLEM FOR THE GENERAL HYPERBOLIC EQUATION

2.3.1. Setting of the problem, reformulation in terms of a system

As before, we consider equation (2.2.1) in the $(n+1)$-dimensional space with the coordinates z, x; $x = (x_1, \dots, x_n)$, $z \in (-\infty, \infty)$, $x \in \mathbb{R}^n$. The coefficients of the equation are assumed to depend only on z. For $z < 0$, the equation coincides with the wave equation

$$u_{tt} = c_0^2(u_{zz} + \Delta_x u).$$

In other words, $a_{00} = a_{11} = \dots = a_{nn} = c_0^2|_{z<0}$, where $c_0 > 0$ is a constant; all the other coefficients of (2.2.1) vanish for $z < 0$. The medium is initially in rest, and it is disturbed by a plane wave incident from the homogeneous half-space $z < 0$:

$$u_{\text{inc}} = \varepsilon(t - (\xi, x) - z\sqrt{1/c_0^2 - |\xi|^2}). \tag{2.3.1}$$

As in Subsection 2.1.5, the field generated by this wave is described by a function dependent on the two variables z and $p = t - (\xi, x)$. We assume that the vector ξ which characterizes the angle of incidence of the wave (2.3.1) is small enough to guarantee no total internal reflection in the medium (the exact formulation of this condition is that the function $\chi(y)$ introduced below must be positive for all y). For $z < 0$, the field generated by the wave (2.3.1) has the form

$$u|_{z<0} = u_{\text{inc}} + u_{\text{refl}},$$

where

$$u_{\text{refl}} = f(\xi, t - (\xi, x) + z\sqrt{1/c_0^2 - |\xi|^2})$$

is the field of the reflected wave. We consider the problem of recovering the coefficients of the equation from a sufficiently large collection of reflected waves corresponding to different angles of incidence (different values of the vector ξ).

Like the problem of Section 2.2, this problem admits of more than one solution. Let us find out the reason of this fact. Since u depends on x and t insofar as it depends on $p = t - (\xi, x)$, we find the differential equation for the function u:

$$\Big(1 - \sum a_{jl}\xi_j\xi_l\Big)u_{pp} = a_{00}u_{zz} - 2\sum a_{0j}\xi_j u_{zp} + b_0 u_z - \sum b_j\xi_j u_p + cu.$$

This equation may be simplified by a change of variables. Introduce a function χ and coordinates y, s:

$$\chi = a_{00}^{-2}\Big(a_{00} - a_{00}\sum a_{jl}\xi_j\xi_l + \Big(\sum a_{0j}\xi_j\Big)^2\Big),$$

$$y = \int_0^z \sqrt{\chi}\,\mathrm{d}z, \qquad s = p + \int_0^z a_{00}^{-1}\sum a_{0j}\xi_j\,\mathrm{d}z. \tag{2.3.2}$$

Upon changing to the variables y, s, the equation takes the form

$$u_{ss} - u_{yy} = \lambda u_y + \mu u_s + \nu u, \tag{2.3.3}$$

where

$$\lambda = \frac{1}{2\chi}\frac{\partial\chi}{\partial y} - \frac{b_0}{a_{00}\sqrt{\chi}},$$

$$\mu = \frac{1}{\sqrt{\chi}}\frac{\partial}{\partial y}\Big(\frac{\sum a_{0j}\xi_j}{a_{00}}\Big) + \frac{b_0\sum a_{0j}\xi_j - a_{00}\sum b_j\xi_j}{a_0^2\chi}, \qquad \nu = \frac{c}{a_{00}\chi}.$$

Finally, we introduce a pair of functions Ψ_1, Ψ_2:

$$\Psi_1 = (u_s - u_y + mu)\exp\Big(\frac{1}{2}\int_0^y(\lambda - \mu)\,\mathrm{d}y\Big),$$

$$\Psi_2 = (u_s + u_y - mu)\exp\Big(\frac{1}{2}\int_0^y(\lambda + \mu)\,\mathrm{d}y\Big),$$

where m is a solution of the differential equation

$$m'_y + \nu + \lambda m + m^2 = 0 \tag{2.3.4}$$

satisfying the condition[6]

$$m|_{y<0} \equiv 0.$$

[6]It is easy to see that the functions λ, μ, ν, for $z < 0$ (or, which is the same, for $y < 0$), are equal to zero; the mentioned initial condition for m is feasible.

It is easy to show that the functions Ψ_1, Ψ_2 satisfy the system

$$\begin{cases} \Psi_{1s} + \Psi_{1y} = q_1 \Psi_2, \\ \Psi_{2s} - \Psi_{2y} = q_2 \Psi_1, \end{cases} \tag{2.3.5}$$

where

$$q_1 = \left(\frac{\mu + \lambda}{2} + m\right) \exp\left(-\int_0^y \mu \, dy\right), \tag{2.3.6}$$

$$q_2 = \left(\frac{\mu - \lambda}{2} - m\right) \exp\left(\int_0^y \mu \, dy\right). \tag{2.3.7}$$

2.3.2. Reduction to an inverse problem investigated above

All the functions introduced above depend on the parameter ξ. Consider the dependence in the case when ξ changes its sign. Evidently, $\chi(-\xi, z) = \chi(\xi, z)$; therefore, $y(-\xi, z) = y(\xi, z)$. Thus, on changing the sign of ξ, the properties of all the functions are the same in z and y: evenness and oddness of the functions with respect to ξ are retained if we go from the variable z to the variable y. So, we see that

$$\lambda(-\xi) = \lambda(\xi), \qquad \mu(-\xi) = -\mu(\xi), \qquad \nu(-\xi) = \nu(\xi),$$

$$m(-\xi) = m(\xi), \qquad q_1(-\xi) = -q_2(\xi), \qquad q_2(-\xi) = -q_1(\xi).$$

Let us consider two scattering processes corresponding to two symmetrically incident waves: the first is incident at the angle determined by the vector ξ (in this case, we denote the solution of the direct problem expressed in terms of the solution of system (2.3.5) by $\vec{\Psi}^+ = (\Psi_1^+, \Psi_2^+))$; and the second, by the vector $(-\xi)$. Introduce a vector function

$$\vec{\Psi}^- = (\Psi_2(-\xi, y, -s), \Psi_1(-\xi, y, -s)).$$

The function $\vec{\Psi}(-\xi, y, s) = (\Psi_1(-\xi, y, s), \Psi_2(-\xi, y, s))$ is a solution of the system

$$\Psi_{1s} + \Psi_{1y} = q_1(-\xi, y)\Psi_2 = -q_2 \Psi_2,$$

$$\Psi_{2s} - \Psi_{2y} = q_2(-\xi, y)\Psi_1 = -q_1 \Psi_1.$$

The q_1 and q_2 on the right are abbreviations of $q_1(\xi, y)$ and $q_2(\xi, y)$, respectively. Hence, $\vec{\Psi}^-$ satisfies system (2.3.5). The following properties of the vectors $\vec{\Psi}^\pm$ are also evident:

$$\vec{\Psi}^+|_{s<y} = 0, \qquad \vec{\Psi}^-|_{s>-y} = 0,$$

$$\vec{\Psi}^+\big|_{y=0} = \begin{pmatrix} \delta(s) \\ h^+(s) \end{pmatrix}, \qquad \vec{\Psi}^-\big|_{y=0} = \begin{pmatrix} h^-(s) \\ \delta(s) \end{pmatrix},$$

where

$$h^+(s) = f'(\xi, s), \qquad h^-(s) = f'(-\xi, -s).$$

Thus we arrive at the inverse problem considered in Section 1.9. Solving this problem, we obtain the functions $q_1(y)$ and $q_2(y)$.

Suppose that the functions q_1 and q_2 have been obtained. We denote by $\tilde{m}$ the function $\tilde{m} = m + \lambda/2$. The functions q_1 and q_2 are expressed in terms of μ and $\tilde{m}$ as follows:

$$q_1 = \Big(\frac{\mu}{2} + \tilde{m}\Big) \exp\Big(-\int_0^y \mu \, \mathrm{d}y\Big),$$

$$q_2 = \Big(\frac{\mu}{2} - \tilde{m}\Big) \exp\Big(\int_0^y \mu \, \mathrm{d}y\Big).$$

Since the functions q_1 and q_2 are known, these equalities constitute a system of equations relative to μ and $\tilde{m}$. We eliminate $\tilde{m}$ from this system. Denoting

$$\exp\Big(\int_0^y \mu \, \mathrm{d}y\Big) \stackrel{\text{def}}{=} \zeta(y)$$

and taking into account that $\mu = \zeta'/\zeta$, we obtain the differential equation (the Riccati equation) for ζ:

$$\zeta' = q_1\zeta^2 + q_2$$

with the evident condition

$$\zeta\big|_{y=0} = 1.$$

From ζ we find μ and $\tilde{m} = q_1\zeta - \zeta'/(2\zeta)$. Consider now equation (2.3.4), which, taking into account the equality $\tilde{m} = m + \lambda/2$, can be rewritten as

$$\tilde{m}' + \tilde{m}^2 = \lambda'/2 + \lambda^2/4 - \nu \stackrel{\text{def}}{=} \Gamma.$$

Thus, if we know reflected waves for the case of two symmetrically incident waves, we can recover the coefficient μ and the combination Γ of the coefficients of equation (2.3.3). However, in the following, we shall often prefer dealing with $\tilde{m}$ instead of Γ.

Remark. Worthy of mention is the case when $\xi = 0$, i.e., when the wave is normally incident. In this case, $\mu \equiv 0$, $q_1 = -q_2 = \tilde{m}$. The function $\tilde{m}$ can be obtained by solving more simple one-dimensional inverse problem (1.1.12)–(1.1.14).

2.3.3. The maximum possible information on the coefficients of equation (2.2.1)

We assume that the one-dimensional inverse problem for equation (2.3.5) has been solved for a sufficiently large collection of values $\pm\xi^{(0)}$, $\pm\xi^{(1)}$, ..., $\pm\xi^{(N)}$. As a result, we find the collection of pairs of functions

$$\mu^{(0)},\ \tilde{m}^{(0)},\quad \mu^{(1)},\ \tilde{m}^{(1)},\quad \ldots,\quad \mu^{(N)},\ \tilde{m}^{(N)}.$$

Each of the pairs of functions depends on its own variable $y^{(0)}, y^{(1)}, \ldots, y^{(N)}$. Since the change of variable (2.3.2) depends on the choice of ξ, the variables $y^{(j)}$ differ. Moreover, the relations between them are unknown: as follows from (2.3.2),

$$\frac{\mathrm{d}y^{(j)}}{\mathrm{d}y^{(i)}} = \sqrt{\frac{\chi^{(j)}}{\chi^{(i)}}}, \tag{2.3.8}$$

but the functions $\chi^{(i)} = \chi(\xi^{(i)}, y^{(i)})$ are unknown. Nevertheless, we can show[7] that there holds the inequality

$$\sqrt{\frac{\chi^{(j)}(y^{(j)})}{\chi^{(i)}(y^{(i)})}} = \exp\left(2\int_0^{y^{(j)}} \tilde{m}^{(j)}(\eta)\,\mathrm{d}\eta - 2\int_0^{y^{(i)}} \tilde{m}^{(i)}(\eta)\,\mathrm{d}\eta\right)\sqrt{\frac{1-c_0^2|\xi^{(j)}|^2}{1-c_0^2|\xi^{(i)}|^2}}.$$

Here $y^{(j)}$ and $y^{(i)}$ are coordinates corresponding to the same depth z. Substituting the last equality into (2.3.8), we obtain a first-order differential equation connecting $y^{(j)}$ and $y^{(i)}$. This is an equation with separable variables; it can be solved in quadratures. Thereby, we can obtain all the variables $y^{(j)}$ $(j = 1, 2, \ldots, N)$ as functions of the same coordinate $y^{(0)}$. Thus, we find all the functions

$$\frac{\chi^{(j)}}{\chi^{(0)}} = \frac{1-\sum a^*_{il}\xi_i^{(j)}\xi_l^{(j)}}{1-\sum a^*_{il}\xi_i^{(0)}\xi_l^{(0)}} \tag{2.3.9}$$

as functions of the argument $y^{(0)}$. The last fact is essential. Indeed, if we know $\chi^{(j)}/\chi^{(0)}$ as a function of $y^{(0)}$, then its dependence on $\xi^{(j)}$ is rather simple: for fixed $\xi^{(0)}$ and $y^{(0)}$, it is a constant minus a quadratic form. If we use the coordinate $y^{(j)}$, this dependence becomes more complicated. The coefficients a^*_{il} of the quadratic form in (2.3.9) are as follows:

$$a^*_{il} = a_{il} - \frac{a_{0i}a_{0l}}{a_{00}}.$$

[7]We omit the proof of this fact because it is rather tedious and not so interesting.

It is easy to show that if we know $\chi^{(j)}/\chi^{(i)}$ for a sufficiently large collection of the vectors $\xi^{(j)}$, we can find all a^*_{il} $(i, l = 1, 2, \dots, n)$. Knowing

$$\mu^{(j)} = \frac{1}{\sqrt{\chi^{(j)}}}\frac{\partial}{\partial y^{(j)}}\Big(\frac{\sum a_{0i}\xi_i^{(j)}}{a_{00}}\Big) + \frac{b_0\sum a_{0i}\xi_i^{(j)} - a_{00}\sum b_i b_i^{(j)}}{a_{00}^2\chi^{(j)}}$$

$$= \frac{\chi^{(0)}}{\chi^{(j)}}\sum a_i^*\xi_i^{(j)},$$

where

$$a_i^* = \Big(\frac{\partial}{\partial z}\Big(\frac{a_{0i}}{a_{00}}\Big) + \frac{b_0}{a_{00}^2}a_{0i} - \frac{b_i}{a_{00}}\Big)\frac{1}{\chi^{(0)}}, \qquad i = 1, 2, \dots, n,$$

for n linearly independent vectors $\xi^{(j)}$, we can derive all the coefficients a_i^* as functions of $y^{(0)}$. Finally, it only remains for us to obtain

$$\Gamma^{(0)} = \frac{\mathrm{d}}{\mathrm{d}y^{(0)}}\tilde{m}^{(0)} + (\tilde{m}^{(0)})^2.$$

The function $\Gamma^{(0)}$ is expressed in terms of the coefficients of the initial equation (2.2.1) as follows:

$$\Gamma^{(0)} = \frac{\gamma}{\chi^{(0)}} + (\chi^{(0)})^{-1/4}\frac{\mathrm{d}^2}{\mathrm{d}y^{(0)2}}(\chi^{(0)})^{1/4},$$

where

$$\gamma = \frac{1}{2}\frac{\mathrm{d}}{\mathrm{d}z}\Big(\frac{b_0}{a_{00}}\Big) + \frac{1}{4}\Big(\frac{b_0}{a_{00}}\Big)^2 - \frac{c_0}{a_{00}}.$$

Thus, we can find the $n(n+1)/2$ coefficients a^*_{il}, the n coefficients a_i^*, and the function $\Gamma^{(0)}$, that is, all together, $(n+1)(n+2)/2$ functions. Note that the number of the function we can obtain coincides with that in the problem of Section 2.2. It is easy to verify that if $\xi^{(0)} = 0$, then the information on the coefficients which can be obtained in these problems is the same.

The problems considered in Sections 2.2 and 2.3 are also investigated in Blagoveshchenskii (1971), [41].

Chapter 3.

Inverse problems for vector wave processes

3.1. INVERSE PROBLEM FOR ELASTICITY EQUATION

3.1.1. General remarks

In this chapter we consider inverse problems for physically important systems of hyperbolic equations such as the elasticity equations in layered inhomogeneous media and the acoustic equations for moving media. Some results are also obtained for hyperbolic systems of general form. This section is devoted to the theory of elastic waves in layered media. The basic results of this section were presented by Blagoveshchenskii (1966).

The inverse Lamb problem will be investigated below.

3.1.2. Setting of the problem

We consider a system of elasticity equations in the half-space with the Cartesian coordinates $z > 0$, $x \in \mathbb{R}^2$. We assume the medium parameters, namely, the density ρ and the Lamé parameters λ and μ, to be functions of z. The displacement vector is taken in the form $\vec{u} = (u, \vec{v})$; where u is the vertical displacement component, and $\vec{v}$ is the two-dimensional vector of horizontal displacement. The elasticity equations are as follows:

$$\rho u_{tt} = \frac{\partial}{\partial z}\Big((\lambda + 2\mu)\frac{\partial u}{\partial z}\Big) + (\lambda + \mu)\frac{\partial}{\partial z}(\nabla, \vec{v}) + \mu\Delta u + \lambda'_z(\nabla, \vec{v}), \quad (3.1.1)$$

$$\rho\vec{v}_{tt} = \frac{\partial}{\partial z}\Big(\mu\frac{\partial\vec{v}}{\partial z}\Big) + (\lambda+\mu)\nabla\Big(\frac{\partial u}{\partial z} + (\nabla,\vec{v})\Big) + \mu'_z\nabla u + \mu\Delta\vec{v}. \qquad (3.1.2)$$

Here ∇ and Δ are the two-dimensional Hamiltonian and Laplace operators,

$$\nabla = \vec{e}_1\frac{\partial}{\partial x_1} + \vec{e}_2\frac{\partial}{\partial x_2}, \qquad \Delta = (\nabla,\nabla);$$

$\vec{e}_1$, $\vec{e}_2$ are the unit vectors. We suppose that the medium is initially at rest, that is, for $t<0$,

$$\vec{u}|_{t<0} = 0. \qquad (3.1.3)$$

Then it begins to vibrate in response to an instantaneous attack directed towards a point on the boundary $z=0$. This is modelled by the boundary conditions

$$\Big((\lambda+2\mu)\frac{\partial u}{\partial z} + \mu(\nabla_x,\vec{v})\Big)\Big|_{z=0} = a\delta(x,t), \qquad (3.1.4)$$

$$\Big(\mu\Big(\frac{\partial\vec{v}}{\partial z} + \nabla_x u\Big)\Big)\Big|_{z=0} = \vec{b}\delta(x,t). \qquad (3.1.5)$$

More precisely, the boundary conditions (3.1.4), (3.1.5) correspond to the case when the medium is disturbed by an impact on the surface. This impact must be directed obliquely to the surface, i.e., it is required that $a\neq 0$, $\vec{b}\neq 0$. The assumption that, say, $\vec{b}=0$ makes the problem essentially more complicated (see Megrabov, 1972).

The inverse problem is to find the coefficients $\lambda(z)$, $\mu(z)$, and $\rho(z)$ from the given functions

$$f(x,t) = u|_{z=0}, \qquad (3.1.6)$$

$$\vec{g}(x,t) = \vec{v}|_{z=0}. \qquad (3.1.7)$$

3.1.3. Solution of the inverse Lamb problem

We make the Fourier transformation with respect to x:

$$\tilde{u} = \int e^{i(k,x)}u(x,z,t)\,dx, \qquad \tilde{\vec{v}} = \int e^{i(k,x)}\vec{v}(x,z,t)\,dx.$$

In terms of the Fourier transform, the problem (3.1.1)–(3.1.7) takes the form

$$\rho\tilde{u}_{tt} = \frac{\partial}{\partial z}\Big((\lambda+2\mu)\frac{\partial\tilde{u}}{\partial z}\Big) - i(\lambda+\mu)\frac{\partial}{\partial z}(k,\tilde{\vec{v}}) - i\lambda'_z(k,\tilde{\vec{v}}), \qquad (3.1.8)$$

$$\rho\tilde{\vec{v}}_{tt} = \frac{\partial}{\partial z}\Big(\mu\frac{\partial\tilde{\vec{v}}}{\partial z}\Big) - ik(\lambda+\mu)\tilde{u}_z - (\lambda+\mu)k(k,\tilde{\vec{v}}) - \mu|k|^2\tilde{\vec{v}} - k\mu_z\tilde{u}, \qquad (3.1.9)$$

$$\tilde{\vec{u}}|_{t<0} \equiv 0 \qquad (\tilde{\vec{u}} = (\tilde{u}, \tilde{\vec{v}})), \tag{3.1.10}$$

$$\Big((\lambda + 2\mu)\frac{\partial \tilde{u}}{\partial z} - \mathrm{i}\mu(k, \tilde{\vec{v}})\Big)\Big|_{z=0} = a\delta(t), \tag{3.1.11}$$

$$\Big(\mu\frac{\partial \tilde{\vec{v}}}{\partial z} - \mathrm{i}k\mu u\Big)\Big|_{z=0} = \vec{b}\,\delta(t), \tag{3.1.12}$$

$$\tilde{u}|_{z=0} = \tilde{f}(k, t), \tag{3.1.13}$$

$$\tilde{\vec{v}}\,|_{z=0} = \tilde{\vec{g}}(k, t). \tag{3.1.14}$$

Here $\tilde{f}$ and $\tilde{\vec{g}}$ are the Fourier transforms of the functions f and $\vec{g}$, respectively.

Setting $k = 0$ in (3.1.8) and in the first components of (3.1.10), (3.1.11), and (3.1.13) results in the following one-dimensional inverse problem:

$$\rho\tilde{u}^0_{tt} = \frac{\partial}{\partial z}\Big((\lambda + 2\mu)\frac{\partial \tilde{u}^0}{\partial z}\Big),$$

$$\tilde{u}^0|_{t<0} = 0,$$

$$(\lambda + 2\mu)\frac{\partial \tilde{u}^0}{\partial z}\Big|_{z=0} = a\delta(t),$$

$$\tilde{u}^0|_{z=0} = \tilde{f}^0(t),$$

where $\tilde{u}^0 = \tilde{u}|_{k=0}$, $\tilde{f}^0(t) = \tilde{f}|_{k=0}$. We have obtained the usual inverse problem investigated in detail in Section 1.6. Solving this problem, we find the function

$$\sigma_l(y_l) = \sqrt{\rho(\lambda + 2\mu)} \tag{3.1.15}$$

as a function of y_l, where

$$y_l = \int_0^z \sqrt{\rho(\lambda + 2\mu)^{-1}}\,\mathrm{d}z \tag{3.1.16}$$

is the time it takes for the longitudinal wave to get to the depth z.

Consider scalar product of equations (3.1.9), (3.1.12), and (3.1.14) by an arbitrary constant vector $\vec{c} \in \mathbb{R}^2$ $((\vec{b}, \vec{c}) \neq 0)$ and set $k = 0$ to obtain the problem

$$\rho\tilde{v}^c_{tt} = \frac{\partial}{\partial z}\Big(\mu\frac{\partial \tilde{v}^c}{\partial z}\Big),$$

$$\tilde{v}^c|_{t<0} = 0,$$

$$\mu\frac{\partial \tilde{v}^c}{\partial z}\Big|_{z=0} = (\vec{b}, \vec{c})\delta(t),$$

$$\tilde{v}^c|_{z=0} = (\tilde{\vec{g}}, \vec{c}),$$

where $\tilde{v}^c$ stands for $(\tilde{\vec{v}}, \vec{c})$. Once again, we have obtained, in fact, the same one-dimensional inverse problem as in the previous case (of course, with other data). Solving this problem yields

$$\sigma_t(y_t) = \sqrt{\rho/\mu} \tag{3.1.17}$$

as a function of y_t,

$$y_t = \int_0^z \sqrt{\rho/\mu}\,\mathrm{d}z, \tag{3.1.18}$$

which is the time it takes for the transverse wave to reach the depth z. Knowing both the functions $\sigma_l(y_l)$ and $\sigma_t(y_t)$, we can easily obtain a relation between y_l and y_t. Really, from (3.1.16) and (3.1.18) it follows that

$$\rho\,\mathrm{d}z = \sqrt{\rho\mu}\,\mathrm{d}y_t = \sqrt{\rho(\lambda + 2\mu)}\,\mathrm{d}y_l,$$

whence

$$\sigma_t(y_t)\,\mathrm{d}y_t = \sigma_l(y_l)\,\mathrm{d}y_l$$

and

$$\int_0^{y_t} \sigma_t(\eta)\,\mathrm{d}\eta = \int_0^{y_l} \sigma_l(\eta)\,\mathrm{d}\eta.$$

Now we fix a vector k such that k and $\vec{b}$ are linearly independent. Let $k_\perp$ be the vector $(k_2, -k_1)$. Evidently,

$$(k_\perp, k) = 0, \qquad (k_\perp, \vec{b}) \neq 0.$$

We introduce a function $w = (k_\perp, \tilde{\vec{v}})$. Then (3.1.9), (3.1.10), (3.1.12), (3.1.14) imply the following one-dimensional inverse problem:

$$\rho w_{tt} = \frac{\partial}{\partial z}\left(\mu \frac{\partial w}{\partial z}\right) - k^2 \mu w,$$

$$w|_{t<0} = 0,$$

$$\mu \frac{\partial w}{\partial z}\bigg|_{z=0} = (k_\perp, \vec{b})\delta(t),$$

$$w|_{z=0} = h(t),$$

where $h(t) = (k_\perp, \tilde{\vec{g}})$. This inverse problem coincides with that considered in Section 1.6. Besides, the function $\sigma_t(y_t)$ has already been found in the previous step. Therefore, solving the last problem, we can find the relation μ/ρ and, hence, the functions

$$\mu = \sigma_t(\mu/\rho)^{1/2}, \qquad \rho = \sigma_t(\mu/\rho)^{-1/2}, \qquad \lambda = \rho^{-1}(\sigma_l^2 - 2\sigma_t^2)$$

(all the above functions are treated as functions of the argument y_t). Finally, y_t and z are related by the evident equality

$$z = \int_0^{y_t} \sqrt{\mu/\rho}\, \mathrm{d}\eta$$

which follows from (3.1.18).

3.2. INVERSE PROBLEM OF SOUND PROPAGATION IN A MOVING LAYERED MEDIUM

3.2.1. Acoustic equations ina moving medium

Sections 3.2 and 3.3 are concerned with inverse problems arising in the theory of propagation of sound waves in a moving medium. The results presented in these sections were published in Nizhnik (1991).

We shall use the system of equations proposed by Blokhintsev (1981), namely:

$$\frac{\partial \rho_1}{\partial t} + (\nabla, U\rho + \rho_1 V) = 0,$$
$$\frac{\partial U}{\partial t} + (V,\nabla)U + (U,\nabla)V + \frac{\nabla p_1}{\rho} - \frac{\rho_1}{\rho^2}\nabla p = 0,$$
$$\frac{\partial S_1}{\partial t} + (V,\nabla)S_1 + (U,\nabla)S = 0. \tag{3.2.1}$$

Here ρ, p, S, V are the density, pressure, entropy, and the vector of flow velocity in the medium, respectively; ρ_1, p_1, S_1, U are the acoustic density, pressure, entropy, and velocity, i.e., small additions to ρ, p, S, and V. System (3.2.1) should be completed by the equation of the medium state $p = \Phi(\rho, S)$. This is the equation for the nondisturbed medium. For the disturbed medium we shall write it in the linearized form

$$p_1 = \frac{\partial \Phi}{\partial \rho}\rho_1 + \frac{\partial \Phi}{\partial S}S_1.$$

Using the notations

$$c^2 = \frac{\partial \Phi}{\partial \rho}, \qquad h = \frac{\partial \Phi}{\partial S},$$

we can rewrite the state equation in the form

$$p_1 = c^2 \rho_1 + h S_1.$$

Besides, the formula

$$\nabla p = c^2 \nabla \rho + h \nabla S$$

evidently holds. Substituting these two formulas in the second equation of system (3.2.1), we eliminate both the pressures p and p_1 from the system to obtain

$$\frac{\partial U}{\partial t} + (V, \nabla)U + (U, \nabla)V + \rho'(c^2 \rho_1 + h S_1) - \frac{\rho_1}{\rho^2}(c^2 \nabla \rho + h \nabla S) = 0. \quad (3.2.2)$$

3.2.2. Transformation of the system for the layered medium

Let z, x_1, x_2 be Cartesian coordinates. We suppose that all the functions characterizing the medium, namely, the functions ρ, S, V, c, h, depend only on z, and that V is always orthogonal to the z-axis. It will be convenient for us to write down three-dimensional vectors in the form $U = \{w, \vec{u}\}$, where $\vec{u}$ is the two-dimensional vector orthogonal to the z-axis, and w is the projection of U along the z-axis. Then the velocity vector has the form $V = \{0, \vec{v}\}$. We consider, for definiteness, the problem of scattering of a plane wave. In this case the acoustic fields are described by the functions ρ_1, w, $\vec{u}$, S_1 dependent only on the variables $s = t - (\xi, x)$ and z. Here ξ is a two-dimensional parameter, $\xi \in \mathbb{R}^2$.

We formulate now the basic result. Suppose that we have a collection of five vectors $\xi^1, \dots, \xi^5$ characterizing the direction of the plane wave incident from the homogeneous half-space being at rest. This collection should satisfy some additional conditions specified below. For each vector ξ^i, a function

$$f^i(t - (\xi^i, x) + (z/c)\sqrt{1 - |\xi^i|^2 c^2}) = w_{\text{refl}}$$

is given which determines the value of the component w of the reflected wave corresponding to the incident wave

$$w_{\text{inc}} = \delta(t - (\xi^i, x) + (z/c)\sqrt{1 - |\xi^i|^2 c^2}).$$

Suppose that we know the values of $\vec{v}$, $\vec{v}'$, c, c', ρ, ρ', and B for $z = 0$, where $B = hS'$. Here we denote by the prime the derivative with respect to z.

These data allow us to recover the collection of the functions $\vec{v}$, c, ρ, and B up to a finite depth[1].

Taking into account all the above assumptions, we can rewrite system (3.2.1) as follows:

$$\begin{gathered}(1-(\vec{v},\xi))\rho_{1s}-\rho(\xi,\vec{u}_s)+(\rho w)_z=0,\\ (1-(\vec{v},\xi))w_s+\rho^{-1}(c^2\rho_1+hS_1)_z-\rho^{-2}(c^2\rho'+hS')\rho_1=0,\\ (1-(\vec{v},\xi))\vec{u}_s+w\vec{v}_z-\xi\rho^{-1}(c^2\rho_{1s}+hS_{1s})=0,\\ (1-(\vec{v},\xi))S_{1s}+wS_z=0.\end{gathered}\tag{3.2.3}$$

Let

$$A=1-(\vec{v},\xi),\qquad \Gamma=A^2-|\xi|^2c^2.$$

The vector ξ is supposed to be small enough to provide $A>0$ and $\Gamma>0$ for all z. The function $\vec{u}$ can be eliminated from the system. To this end, we sum the first equation multiplied by A and the third equation multiplied scalarly by $\rho\xi$. Then we arrive at the following system:

$$\begin{gathered}\rho_{1s}+\Gamma^{-1}A(\rho w)_z+\Gamma^{-1}\rho wA_z-\Gamma^{-1}|\xi|^2hS_{1s}=0,\\ w_s+A^{-1}\rho^{-1}(c^2\rho_1+hS_1)_z-A^{-1}\rho^{-2}(c^2\rho_z-hS_z)\rho_1=0,\\ S_{1s}=-A^{-1}S_zw.\end{gathered}\tag{3.2.4}$$

Analogously, we can eliminate from system (3.2.4) the functions ρ_1 and S_1 (for this purpose, it suffices to differentiate the second equation of system (3.2.4) with respect to s and then replace ρ_{1s} and S_{1s} by their expressions from the first and third equations).

As a result, we obtain one second-order equation for w:

$$\begin{aligned}w_{ss}&+A^{-1}\rho^{-1}\frac{\partial}{\partial z}(-\Gamma^{-1}Ac^2(\rho w)'_z+\Gamma^{-1}\rho c^2A'w-\Gamma^{-1}BAw)\\ &+A^{-2}\rho^{-2}\Gamma^{-1}(c^2\rho'+B)(A^2(\rho w)'_z-\rho AA'w+|\xi|^2Bw)=0.\end{aligned}\tag{3.2.5}$$

We simplify equation (3.2.5) using the change of variables:

$$y=\int_0^z c^{-1}\sqrt{\Gamma}\,d\xi,\qquad w=\Gamma^{1/4}(\rho c)^{-1/2}\Psi.$$

[1]The functions h and S can not be found from the functions f^i separately; only their combination $B=hS'$ can be found.

As a result, we get the equation for Ψ:

$$\Psi_{ss} - \Psi_{yy} + Q\Psi = 0, \tag{3.2.6}$$

where

$$Q = \frac{c^2}{\Gamma}n' + n^2 + \frac{1}{2}\Big(\frac{c^2}{\Gamma}\Big)' n - \frac{rc^2}{A^2}, \qquad n = (\ln(\Gamma^{-1/4}A))' - \lambda,$$

$$\lambda = \frac{1}{2}(\ln(\rho/c))' + \nu, \qquad \nu = \frac{B}{\rho c^2}, \qquad r = \nu((\rho'/\rho) - \nu). \tag{3.2.7}$$

It will be more convenient for us to consider the first relation of (3.2.7) in another form, namely,

$$\begin{aligned}(\vec{v}'', \xi)L &+ \frac{c|\xi|^2}{2\Gamma}c'' - \lambda' - \frac{\Gamma}{A^2}r \\ &= \frac{\Gamma}{c^2}Q - n\Big(\ln\frac{c}{\sqrt{\Gamma}}\Big)' - n^2 + \Big(\frac{A'}{A}\Big)^2 - \frac{1}{4}\Big(\frac{\Gamma'}{\Gamma}\Big)^2 + \frac{A'^2 - |\xi|^2 c'^2}{2\Gamma} \\ &= F(c, c', \vec{v}, \vec{v}', \lambda, y).\end{aligned} \tag{3.2.8}$$

Here $L = A^{-1} - A/(2\Gamma)$. If we know the function Ψ for $y < 0$,

$$\Psi|_{y<0} = (\rho_0 c_0)^{1/2}(1 - |\xi|^2 c_0^2)^{-1/4}(\delta(s - y) + f(s + y)),$$

then by solving the one-dimensional inverse problem considered in Subsection 1.6.3, we find the function $Q(y)$.

Suppose now that we have found the function $Q(y)$ for five different values $\xi = \xi^i$. Each function depending on the choice of ξ and corresponding to the value $\xi = \xi^i$ will be written with the superscript i. Since the change of the variables $z \to y$ depends on the choice of ξ (we recall that Γ depends on ξ), each of the functions Q^i is a function of its variable y^i.

Now, we consider a 5×5 matrix whose i-th row is

$$\Big(L^i\xi_1^i,\ L^i\xi_2^i,\ \frac{c|\xi^i|^2}{2\Gamma^i},\ -1,\ \frac{\Gamma^i}{(A^i)^2}\Big).$$

Here ξ_1^i, ξ_2^i are the components of the vector ξ^i,

$$A^i = 1 - (\vec{v}, \xi^i), \quad \Gamma^i = (A^i)^2 - |\xi^i|^2 c^2, \quad L^i = (A^i)^{-1} - \frac{A^i}{2\Gamma^i}.$$

We denote by $\Delta(z)$ the determinant of this matrix. Let

$$\Delta(0) \neq 0. \tag{3.2.9}$$

Since $\Delta(z)$ depends on the functions $\vec{v}$, c and on the choice of the vectors ξ^i, condition (3.2.9) should be met when choosing the vectors ξ^i for fixed $\vec{v}(0)$ and c.

Introduce now a vector function $X(z) \in \mathbb{R}^{12}$ whose components are defined by the formulas

$$x_i = y^i(z), \quad i = 1, 2, \dots, 5, \quad x_6 = v_1(z), \quad x_7 = v_2(z),$$

$$x_8 = c(z), \quad x_9 = v_1'(z), \quad x_{10} = v_2'(z), \quad x_{11} = c'(z), \quad x_{12} = \lambda(z).$$

Note that five relations (3.2.8) written for each $\xi = \xi^i$ form a linear algebraic system relative to v_1'', v_2'', c'', λ', and r such that its right-hand side involves known functions of X. The matrix of this system if the 5×5 matrix introduced above. Its elements are functions of X (more precisely, of x_6, x_7, x_8). Because of $\Delta(0) \neq 0$, the continuity implies that there exists $z_0 > 0$ such that $\Delta(z) \neq 0$ for $0 \leq z \leq z_0$. Therefore, inverting this matrix, we find (taking into account that $v_1'' = x_9'$, $v_2'' = x_{10}'$, $c'' = x_{11}'$, $\lambda' = x_{12}'$)

$$r = H_0(X), \tag{3.2.10}$$

$$x_k' = H_k(X) \tag{3.2.11}$$

for $k = 9, 10, 11, 12$. Equations similar to (3.2.11) hold for $k = 1, 2, 3, 4, 5$ as well. This follows from the definition of the coordinates y^i

$$x_i' = (y^i)' = c^{-1}\sqrt{\Gamma^i}.$$

Finally, for $k = 6, 7, 8$, equations of the form (3.2.11) follow from the definition of x_9, x_{10}, x_{11}: $x_k' = x_{k+3}$. Thus, (3.2.11) is a system of twelve first-order ordinary differential equations with derivatives in its left-hand side. Initial conditions for X are also known: for $k = 1, 2, 3, 4, 5$, $x_k|_{z=0} = 0$; for $k = 6, 7, 8, 9, 10, 11$, $x_k|_{z=0}$ are the values of v_i, c, v_i', c' for $z = 0$; $x_{12}|_{z=0} = \lambda|_{z=0}$, where λ is the explicit function of ρ, ρ', c, c', and B (see formulas (3.2.7)) whose values are known for $z = 0$. Therefore, the problem of determination of $X(z)$ is the Cauchy problem for system (3.2.11). Having solved this problem, we thus find the functions $\vec{v}(z)$, $c(z)$, their derivatives, and λ. It remains to find ρ and B. To this end, we find r from (3.2.10). Eliminating ν from the relations (see (3.2.7)

$$\lambda = \frac{1}{2}((\ln \rho)' - (\ln c)' + 2\nu), \qquad r = \nu((\ln \rho)' + \nu),$$

we obtain

$$(\ln \rho)'^2 = (2\lambda + (\ln c)')^2 - 4r.$$

Hence, we find $(\ln \rho)'$ to within the sign. The sign is determined by the value $\rho'|_{z=0}$ if $\rho'|_{z=0} \neq 0$ and then from the continuity if ρ' does not vanish. After getting ρ, we find $\nu = \lambda - (1/2)(\ln \rho)' + (1/2)(\ln c)'$ and $B = \rho c^2 \nu$.

The most essential restriction imposed on the inverse problem data seems to be condition (3.2.9). This condition means that $\Delta(0)$ as a function of ξ^i is not equal to zero identically. This follows from the representation of $\Delta(0)$ for small ξ^i:

$$\Delta(0) = \frac{1}{8} c^5 \Delta_0 + O\Big(\Big(\sum_{i=1}^{5} |\xi^i|\Big)^9\Big).$$

Here Δ_0 is a homogeneous eighth-order polynomial of the components of the vectors ξ^i; Δ_0 is independent of $\vec{v}(0)$ and $c(0)$. For example, if $\xi_2^i = 0$ for $i = 1, 2, 3, 4$, then

$$\Delta_0 = \xi_2^5 \Big(\sum_{i=1}^{4} \xi_1^i\Big) \prod_{0<i<k\leq 4} (\xi_1^i - \xi_1^k).$$

At the same time, this representation implies that $\Delta(0)$ may vanish for nontrivial collections of the vectors ξ^i.

3.3. THE CASE OF ONE-DIMENSIONAL SOUND PROPAGATION

3.3.1. General description of the problems in question

Here we consider the one-dimensional case: the functions which characterize the medium (the flow and sound velocities, density, and entropy) depend only on the coordinate z. The flow in the medium is directed parallel to the z-axis. The acoustic waves are propagated in the same direction: the functions describing the acoustic field depend on z and t. This situation takes place, for example, when sound is propagated in tubes along which liquid or gas flows. We suppose that the acoustic field is known for $z = 0$.

These data are not sufficient to recover all the medium characteristics for $z > 0$. Only some of them can be recovered if the rest are given (or if we know some relations between them). Thus, there arise a lot of various physically reasonable settings of the inverse problems. Two of them will be considered in more detail. These are the problems of determination of one of the medium characteristics provided all the others are known. The first problem is to find the medium density, in the second problem we seek for the flow velocity $v(z)$. In both the cases we suggest that $v > 0$ for all z (probing is cocurrent with the flow).

3.3.2. Mathematical setting of the problems

Let the flow velocity in the medium have the form $V = \{v, \vec{0}\}$. The acoustic velocity of the medium particles is $U = \{w, \vec{0}\}$. The acoustic field is described by the triple of the functions ρ_1, w, S dependent only on z and t. Suppose also that ρ, c, v, h, and S are functions of z. System (3.2.1), after eliminating the pressure in accordance with (3.2.2), has the form

$$\frac{\partial \rho_1}{\partial t} + \frac{\partial}{\partial z}(\rho_1 + \rho w) = 0,$$

$$\frac{\partial w}{\partial t} + v\frac{\partial w}{\partial z} + w\frac{\partial v}{\partial z} + \frac{1}{\rho}\frac{\partial}{\partial z}(c^2\rho_1 + hS_1) - \frac{\rho_1}{\rho}(c^2\rho' + hS') = 0,$$

$$\frac{\partial S_1}{\partial t} + v\frac{\partial S_1}{\partial z} + wS_1 = 0. \tag{3.3.1}$$

To rewrite the system, we introduce new functions

$$Y_0 = S_1,$$

$$Y_1 = \frac{v+c}{2\sqrt{\rho c}}\Big(c\rho_1 + \rho w + \frac{h}{c}S_1\Big),$$

$$Y_2 = \frac{v-c}{2\sqrt{\rho c}}\Big(c\rho_1 - \rho w + \frac{h}{c}S_1\Big)$$

and new independent variables

$$y = \int_0^z \frac{c(\zeta)\,d\zeta}{c^2(\zeta) - v^2(\zeta)}, \qquad s = t + \int_0^z \frac{v(\zeta)\,d\zeta}{c^2(\zeta) - v^2(\zeta)}.$$

Then system (3.3.1) becomes

$$\dot{Y} + \Lambda Y' = QY. \tag{3.3.2}$$

Here $Y = Y(y, s)$ is the vector (Y_0, Y_1, Y_2); Λ is the diagonal matrix $\Lambda = \operatorname{diag}[\mu, 1, -1]$; $\mu = v/c$; and Q is the matrix with the elements

$$q_{ii} = 0, \qquad i = 0, 1, 2,$$

$$q_{0k} = -S'c^{-3/2}\rho^{-1/2}(1 - (-1)^k\mu)^{-1}, \qquad k = 1, 2,$$

$$q_{k0} = -\frac{1}{2}\sqrt{\frac{c}{\rho}}\Big(\frac{h}{\rho c^3}(\rho' c^2 + hS') + (-1)^k \Big(\frac{\mu h}{c}\Big)'\Big), \qquad k = 1, 2,$$

$$q_{ik} = (-1)^i \Big(\frac{1}{2}\Big(\ln \frac{\rho}{c}\Big)' + hS'\rho^{-1}c^{-1}(1 + (-1)^i \mu)^{-1}\Big), \qquad i, k = 1, 2 \quad (i \neq k),$$

the dot and prime denote differentiation with respect to s and y, respectively. We suppose that

$$Y|_{s<y} = 0, \qquad Y|_{y=0} = (\varphi_0(s), \delta(s) + \varphi_1(s), \varphi_2(s)),$$

where $\varphi_0(s)$, $\varphi_1(s)$, $\varphi_2(s)$ are known functions (the inverse problem data). For system (3.3.2), the following three velocities of propagation of the disturbances are typical: $(+1)$ and (-1) correspond to the cases when the disturbances move to the right or to the left with the sound speed (the variables s and y were taken so as to equate these velocities to unit); the velocity μ is in the case when the disturbances propagate with the flow speed in the medium.

It will be shown below that the data $\varphi_0(s)$, $\varphi_1(s)$, $\varphi_2(s)$ are insufficient for the unique determination of all the functions involved in the expressions for the coefficients of system (3.3.2), namely:

$$\{\mu, c, \rho, h, S\}. \tag{3.3.3}$$

We can set various inverse problems of determining some of the functions from the collection (3.3.3), considering the remaining functions as known. In the following, we shall suppose that the function μ does not change its sign and $|\mu| < 1$ (the speed of flow is less than that of sound). The two cases should be mentioned: $\mu > 0$ (acoustic probing is "cocurrent with the flow") and $\mu < 0$ (probing is "opposite to the flow"). These cases are essentially different. Really, if $\mu > 0$, the two functions $\varphi_0(s)$ and $\varphi_1(s)$ should be given for the direct problem to be well-posed. To formulate the inverse problem in this case, we set additionally one function $\varphi_2(s)$. Therefore, it is natural to seek for one function from the collection (3.3.3) if the others are known. If $\mu < 0$, then in the direct problem we should set only one function φ_1. Consequently, in the inverse problem, we can assume two functions from (3.3.3) to be desired. Note that modifications of the problems are also of interest, in which either ρ_1 or S_1 is replaced by the pressure p_1 using the state equation

$$p_1 = c^2 \rho_1 + hS_1.$$

We may impose various natural restrictions on the functions of the collection (3.3.3). For example, we may consider the problems in which the state equation $p = \Phi(\rho, S)$ has a fixed form, we may require the equation of continuity $(\rho v)' = 0$ to hold[2], and so on. In each of these cases we may vary the problems choosing different functions of (3.3.3) to be sought for.

From the above discussion it is clear that a great number of physically reasonable problems can be posed. We shall restrict our consideration to two of them in the case when $\mu > 0$.

In Problem 1, we shall seek for the function $\rho > 0$. The functions μ, c, h, S and the constant $\rho_0 = \rho|_{y=0}$ are supposed to be known.

In Problem 2, the functions ρ, c, h, S and the constant $\mu_0 = \mu|_{y=0}$ are given. The desired function is $\mu(y)$.

3.3.3. Formulation of the results

The following statements hold.

A. Suppose that the functions $\varphi_i(s)$ are continuous in the segment $[0, 2y_0]$; the functions v, c, h, and S are continuously differentiable for $y \in [0, y_0]$; $\rho_0 > 0$. Then the solution of Problem 1 is unique in the segment $[0, y_0]$, and it can be found using a nonlinear system of Volterra integral equations. The solution exists if y_0 is sufficiently small.

B. Before formulating the statement, we introduce a class of functions $T_{y_0\alpha\beta M}$. We say that a function $\mu(y)$ belongs to the class $T_{y_0\alpha\beta M}$ if it is differentiable for $0 \le y \le y_0$, it satisfies the estimates $0 < \alpha \le \mu(y) \le \beta < 1$, and μ' satisfies the Lipschitz condition

$$|\mu'(y_1) - \mu'(y_2)| \le M|y_1 - y_2|$$

for each y_1, y_2 from the segment $[0, y_0]$.

We suppose that ρ, c, h, S' do not vanish for $0 \le y \le y_0$; h is continuously differentiable; ρ, c, S are twice continuously differentiable; and ρ'', c'', S'', and h' satisfy the Lipschitz condition. The functions $\varphi_i(s)$ $(i = 0, 1, 2)$ are continuously differentiable, and $\dot{\varphi}_i(s)$ satisfy the Lipschitz condition. Then Problem 2 has at most one solution in the class $T_{y_0\alpha\beta M}$. To find the solution, we construct a system of nonlinear integral equations of a somewhat abnormal type.

3.3.4. Integral equations for Problem 1

Using the methods analogous to those proposed in Section 1.2, it is easy to find the character of singularity of the solution Y: $Y_1 = \delta(s - y) + \tilde{Y}_1$, where

[2] A failure of the equation of continuity has physical meaning, for example, in the case of flow leakage or gas escape.

$\tilde{Y}_1$, as well as Y_0 and Y_2 are classical functions continuous for $s > y$, and

$$Y_0(y,y) = (1-\mu)^{-1} q_{01}, \tag{3.3.4}$$

$$\tilde{Y}_1(y,y) = \varphi_1(0) + \int_0^y \left(q_{10} q_{01} (1-\mu)^{-1} + \frac{1}{2} q_{12} q_{21} \right) d\eta, \tag{3.3.5}$$

$$Y_2(y,y) = \frac{1}{2} q_{21}. \tag{3.3.6}$$

Here, as before, $Y_i(y,y) = \lim_{s \to y+0} Y_i(y,s)$. Also we shall need an expression for $\dot{Y}_2(y,y)$. We can show that this expression involves μ'. Upon rearranging, this expression gets the form:

$$\mu' = \frac{2\rho c^2 (1+\mu)^2 (1-\mu)}{hS'} \Big[\dot{Y}_2(y,y) - \frac{1}{2} q_{21} Y_1(y,y) + \frac{1}{8} \left(\ln \frac{\rho}{c} \right)'' + \frac{1}{4(1+\mu)} \left(\frac{hS'}{\rho c^2} \right)' - \frac{S'}{4(1-\mu^2)\rho c} \left(\frac{h}{\rho c^3} (\rho' c^2 + hS') + \mu \left(\frac{h}{c} \right)' \right) \Big]. \tag{3.3.7}$$

We integrate now each of the equations of system (3.3.2) along the corresponding characteristic of this system. The equations of these characteristics are as follows (see Fig. 3.1):

$$s_1 = s - \tau(\eta, y), \quad s_1 = s + \eta - y, \quad s_1 = s + y - \eta,$$

where

$$\tau(\eta, y) = \int_\eta^y \mu^{-1}(\eta_1)\, d\eta_1$$

is the time it takes for a medium particle to run from the point η to the point y. All the characteristics end at the point y, s of the plane of the variables s_1, η. The first characteristic originates at the point η^*, s^* such that $\eta^* = 0$, $s^* = s - \tau(0,y)$ if $s > \tau(0,y)$. If $s < \tau(0,y)$, then $s^* = \eta^*$, where η^* is the root of the equation $\eta^* = s - \tau(\eta^*, y)$.

After integrating along the characteristics, we obtain

$$Y_0(y,s) = \int_{\eta^*}^y \mu^{-1} (q_{01}(\eta) Y_1(\eta, s - \tau(\eta,y)) + q_{02}(\eta) Y_2(\eta, s - \tau(\eta, y)))\, d\eta + Y_0(\eta^*, s^*), \tag{3.3.8}$$

$$Y_1(y,s) = \int_0^y (q_{10}(\eta) Y_0(\eta, s + \eta - y) + q_{12}(\eta) Y_2(\eta, s + \eta - y))\, d\eta + \varphi_1(s-y), \tag{3.3.9}$$

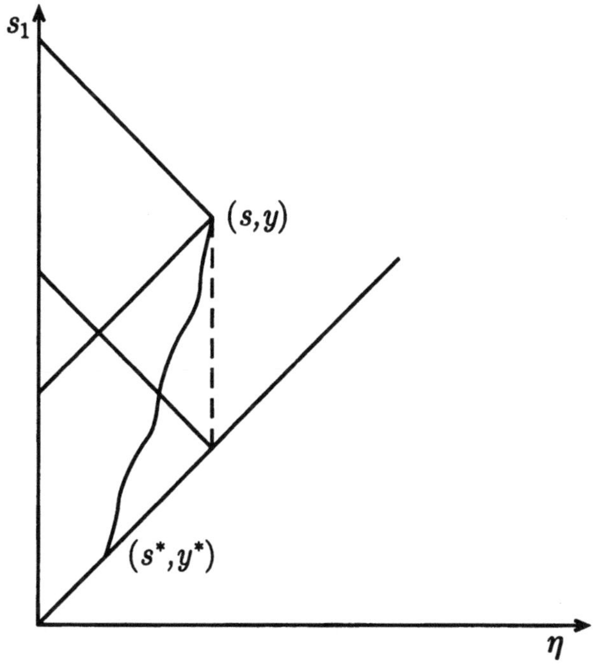

Fig. 3.1

$$Y_2(y,s) = -\int_0^y (q_{20}(\eta)Y_0(\eta, s-\eta+y) + q_{21}(\eta)Y_1(\eta, s-\eta+y))\,d\eta$$
$$+\varphi_2(s+y). \tag{3.3.10}$$

The function $Y_0(\eta^*, s^*)$ in (3.3.8) is assumed to be $\varphi_0(s - \tau(0,y))$ if $s > \tau(0,y)$ or $(1-\mu)^{-1}q_{01}|_{\eta=\eta^*}$ otherwise. We introduce now an unknown function $\gamma = \rho'/\rho$. The coefficients q_{ik} of system (3.3.2) are explicit functions of γ, ρ as well as of the quantities given in Problem 1. The function $\rho(y)$ is everywhere $\rho_0 \exp(\int_0^y \gamma(\eta)\,d\eta)$. From relation (3.3.6) and the definition of q_{21}, we have

$$\gamma = \frac{c'}{c} - \frac{2hS'}{c^2(1+\mu)\rho} + 4Y_2(y,y). \tag{3.3.11}$$

Now we replace $Y_2(y,y)$ in (3.3.11) by its expression from (3.3.10) and $Y_0(\eta^*,\eta^*)$ in (3.3.8), for $s > \tau(0,y)$, by the expression

$$(1-\mu(\eta^*))^{-1}q_{01}(\eta^*) = -\frac{c^{-3/2}(\eta^*)S'(\eta^*)}{1-\mu^2(\eta^*)}\rho^{-1/2}(\eta^*).$$

Equations (3.3.8)–(3.3.11), after the above transformations, form a nonlinear Volterra system of equations of second kind, whence statement A follows.

As is seen, the methods used in this section are analogous to those presented in Section 1.3.

3.3.5. Integral equations for Problem 2

To prove statement B we use, in essence, the same system (3.3.8)–(3.3.11). However, it has to be modified. The main desired function is μ. Note that in the right-hand side of the system (3.3.8)–(3.3.11) μ' occurs under the integral sign. Therefore, it is convenient to take μ' for the desired function. Then the function μ will be as follows:

$$\mu = \mu_0 + \int_0^y \mu'(\eta)\, d\eta.$$

One of the equations of the required system we be equation (3.3.7) with q_{21} and $Y_1(y, y)$ replaced by their expressions. The resulting equation contains $\dot{Y}_2(y, y)$. Therefore, we need to seek for the functions $\dot{Y}_0$, $\dot{Y}_1$, $\dot{Y}_2$ as well. To this end, we differentiate each of equations (3.3.8)–(3.3.10) with respect to s and denote them by (3.3.8a)–(3.3.10a), respectively. These equations differ from (3.3.8)–(3.3.10) only in having $\dot{Y}_i$ instead of Y_i (except for the last term in (3.3.8)) and $\dot{\varphi}_i$ instead of φ_i. In (3.3.8a)

$$\frac{\partial}{\partial s} Y_0(\eta^*, s^*) = \dot{\varphi}_0(s^*)$$

if $s > \tau(0, y)$. If $s < \tau(0, y)$, then it is easy to show that

$$\frac{\partial Y_0(\eta^*, s^*)}{\partial s} = \left(\frac{S'}{c\sqrt{\rho c}}\right)' \frac{\mu}{(1-\mu^2)(1+\mu)} + \frac{2S'\mu\mu'}{c\sqrt{\rho c}(1-\mu)^3(1+\mu)^2}.$$

Finally, we eliminate $\mu'(\eta^*)$ from the last formula using (3.3.7), and then eliminate $\dot{Y}_2(y, y)$ from (3.3.7) substituting its expression from (3.3.10a). As a result, we obtain a system of integral equations for Problem 2 relative to the functions μ', $\dot{Y}_0$, $\dot{Y}_1$, $\dot{Y}_2$. Unfortunately, this system is rather bulky. This is why we do not write it here and only give the way of its construction.

It should be pointed out that the obtained system has a distinguishing feature. Note that equation (3.3.8a) of the system has integrals of the type

$$\int_{\eta^*}^{y} \mu^{-1}(\eta) q_{0i}(\eta) \dot{Y}_i(\eta, s + \tau(\eta, y))\, d\eta.$$

In these integrals the argument of the unknown function $\dot{Y}_i$ involves $\tau(\eta, y)$ which depends, in its turn, on the unknown function μ'. In other words, the integral of the unknown function $\dot{Y}_i$ is taken along an unknown curve.[3] To demonstrate the way of investigation of the system, we shall consider a model problem whose system of equations is much more simple, but it does have the feature mentioned above.

3.3.6. The model problem

We consider the following inverse problem. Suppose that $Y(y,s) = (Y_1, Y_2)$ satisfies the following system

$$\dot{Y}_1 + Y_1' = 0,$$

$$\nu\dot{Y}_2 - Y_2' = -Y_2 + \nu(\nu+1)Y_1 \tag{3.3.12}$$

and the conditions

$$Y|_{s<y} \equiv 0, \tag{3.3.13}$$

$$Y_1|_{y=0} = \delta(s), \qquad Y_2|_{y=0} = \varphi(s). \tag{3.3.14}$$

Here $\nu = \nu(y)$ is an unknown function of y, $\varphi(s)$ is a given function such that

$$|\varphi(s_1) - \varphi(s_2)| \le M|s_1 - s_2|.$$

System (3.3.12) describes the wave process admitting only two alternatives for the velocity of the wave propagation: 1 and $(-1/\nu)$. We do not restrict ν to be either positive or negative (that is, waves may run to any direction). However, we assume that $\nu > -1$; this means that if the wave modelled by the function Y_2 runs to the right, then its speed is greater than unity. Evidently, $Y_1 \equiv \delta(s-y)$. The component Y_2 has the singularity

$$Y_2 \sim \nu\delta(s-y).$$

Hence, we obtain two integral equations for Y_2 and ν, namely, for $s > y$,

$$Y_2(y,s) = \int_0^y Y_2(\eta, s+\tau(\eta,y))\,d\eta + \varphi(s+\tau(0,y)),$$

$$\nu(y) = Y_2(y,y) = \int_0^y Y_2(\eta, y+\tau(\eta,y))\,d\eta + \varphi(y+\tau(0,y)), \tag{3.3.15}$$

[3]This feature of the system (3.3.7), (3.3.8a)–(3.3.10a) is so essential that it is questionable whether it can be called a system of integral equations. However, this question is inherently of nomenclature type. Therefore, we do refer to this system as the system of integral equations.

where $\tau(\eta, y) = \int_\eta^y \nu(\eta_1)\,d\eta_1$. We see that this system of integral equations being rather simple has the main feature which the system (3.3.7), (3.3.8a)–(3.3.10a) has.

Let a number $\nu_0 > -1$ be given. We assume that $\nu < \nu_0$. System (3.3.15) can be written in the form

$$Z = KZ,$$

where Z is a pair of continuous functions $Y_2(y,s)$, $\nu(y)$; $\nu(y)$ is defined for $y \in [0, y_0]$, $y_0 > 0$; $Y_2(y,s)$ is defined for $y < s < (\nu_0+1)y_0 - \nu_0 y$. The form of the operator K is evident. Define the norm of Z by

$$\|Z\| = \max_{\substack{y\in[0,y_0]\\ s\in[y,(\nu_0+1)y_0-\nu_0 y]}} \{|Y_2(y,s)|, |\nu(y)|\}.$$

We consider the class $T_{y_0\nu_0 M}$ of the pairs of continuous functions Y_2, ν such that $-1 < \nu < \nu_0$ and Y_2 satisfies the Lipschitz condition

$$|Y_2(y,s_1) - Y_2(y,s_2)| \le M|s_1 - s_2|$$

with a fixed constant $M > 0$. Suppose that two pairs Z and $\tilde Z$ from the class $T_{y_0\nu_0 M}$ are given. Let us estimate the norm $\|KZ - K\tilde Z\|$. The first component is as follows:

$$\int_0^y (Y_2(\eta, s+\tau(\eta,y)) - \tilde Y_2(\eta, s+\tilde\tau(\eta,y)))\,d\eta + \varphi(s+\tau(0,y)) - \varphi(s+\tilde\tau(0,y)).$$

Its modulus is evidently not greater than

$$\begin{aligned}
&\int_0^y |Y_2(\eta, s+\tau(\eta,y)) - \tilde Y_2(\eta, s+\tilde\tau(\eta,y))|\,d\eta \\
&+ |\varphi(s+\tau(0,y)) - \varphi(s+\tilde\tau(0,y))| \\
&\quad \le \int_0^y \Big(|Y_2(\eta, s+\tau(\eta,y)) - Y_2(\eta, s+\tilde\tau(\eta,y))| \\
&\qquad\qquad + |Y_2(\eta, s+\tilde\tau(\eta,y)) - \tilde Y_2(\eta, s+\tilde\tau(\eta,y))|\Big)\,d\eta \\
&\quad + M|\tau(0,y) - \tilde\tau(0,y)| \\
&\quad \le M\int_0^y |\tau(\eta,y) - \tilde\tau(\eta,y)|\,d\eta + \max|Y_2 - \tilde Y_2| \int_0^y d\eta \\
&\quad + M|\tau(0,y) - \tilde\tau(0,y)|.
\end{aligned}$$

Note that

$$|\tau(\eta, y) - \tilde{\tau}(\eta, y)| \leq \int_{\eta}^{y} |\nu(\eta_1) - \tilde{\nu}(\eta_1)| \,\mathrm{d}\eta_1 \leq (y - \eta) \max |\nu(\eta_1) - \tilde{\nu}(\eta_1)|.$$

Finally, we obtain that the modulus of the first component of $KZ - K\tilde{Z}$ does not exceed

$$\Big(M \int_0^y (y - \eta) \,\mathrm{d}\eta + (y + My)\Big) \|Z - \tilde{Z}\| \leq y\Big(1 + M + \frac{1}{2} My\Big) \|Z - \tilde{Z}\|.$$

The second component of $KZ - K\tilde{Z}$ can be estimated analogously. Taking into account that $y < y_0$, we have

$$\|KZ - K\tilde{Z}\| \leq y_0\Big(1 + M + \frac{1}{2} My_0\Big) \|Z - \tilde{Z}\|.$$

We choose y_0 such that

$$\alpha = y_0\Big(1 + M + \frac{1}{2} My_0\Big) < 1.$$

This means that the operator K is a contraction operator in the class $T_{y_0 \nu_0 M}$. Hence, the solution of equation (3.3.15) is unique (if $Z = KZ$, $\tilde{Z} = K\tilde{Z}$, and $Z, \tilde{Z} \in T_{y_0 \nu_0 M}$, then $\|Z - \tilde{Z}\| = \|KZ - K\tilde{Z}\| \leq \alpha \|Z - \tilde{Z}\| < \|Z - \tilde{Z}\|$, which is impossible if $\|Z - \tilde{Z}\| \neq 0$). The conditional stability can also be easily shown. Really, if $\max_s |\varphi(s) - \tilde{\varphi}(s)|$ is small and the corresponding solutions Z and $\tilde{Z}$ belong to the class $T_{y_0 \nu_0 M}$, then they are close to each other ($\|Z - \tilde{Z}\|$ is small). The system (3.3.7), (3.3.8a)–(3.3.10a) can be considered in a similar way.

3.4. INVERSE PROBLEMS FOR HYPERBOLIC SYSTEMS

3.4.1. General remarks. Setting of a problem

In Sections 3.4 and 3.5 we shall consider a problem of determining coefficients in the lowest terms of hyperbolic systems in the case when waves may travel with different velocities. In Section 3.4 we study a more simple problem: to find some of these coefficients. In Section 3.5 we shall determine all the coefficients in the second-order system.

Let us consider the system

$$u_t + \Lambda u_x - Qu = 0. \qquad (3.4.1)$$

Here $t \in \mathbb{R}$; $x \in (0, l)$, $l \leq \infty$; $u(x,t)$ is a vector function whose values are in $\mathbb{R}^n$ or $\mathbb{C}^n$; Λ, Q are $n \times n$ matrices dependent on x; Λ is a given real diagonal matrix which is continuously dependent on x and invertible. The matrix $Q(x)$ is known partially: some of its elements are unknown. It is required to find the unknown elements[4] of the matrix Q from the measurements of wave fields on the boundary $x = 0$. The wave fields are assumed to be generated by sources located at the point $x = 0$ (or, for scattering problems, by waves coming from the half-space $x < 0$).

The inverse problems for system (3.4.1) were considered by Romanov and Slinucheva (1972), Romanov (1976), Belinskii (1975, 1976a, 1976b). However, the problems considered in these works presuppose the setting of nontrivial initial conditions in the whole interval or, for the multidimensional case, in the whole domain where the coefficients must be determined. In the monograph by Romanov (1987) the inverse problem was considered for the case when the medium is probed from both the ends of the interval. We should also mention the paper by Nizhnik and Tarasov (1977) and the monograph by Nizhnik (1991) devoted to the inverse scattering problem for system (3.4.1) (in these works the coefficients may depend on time). This section follows the monograph by Romanov (1987).

Studying the inverse problems for system (3.4.1) in the case when the wave velocity may take various values (the diagonal elements of the matrix Λ), we met some difficulties. These difficulties are associated with the fact that the integral equations obtained by our methods are not of Volterra type. This obstruction, both in the works by Romanov (1976), Romanov and Slinucheva (1972), Romanov and Belinskii (1976), Belinskii (1975, 1976a, 1976b), and in the present work, is not removed; rather it is got around by setting up the problem in a special form. We overcome the difficulty by assuming that not all the coefficients of Q are unknown. This assumption seems to be more natural. For example, this approach may be applied to inverse problems for the elasticity equation.

3.4.2. Formulation of the direct problem

We suppose that the matrix functions $\Lambda(x)$ and $Q(x)$ are continuous. First we impose restrictions on the structure of the matrix Λ. Suppose that $n = \sum_{i=0}^{2} n_i$, and the numbers $n_1 \geq 0$, $n_0 > 0$, $n_2 > 0$ are independent of x. Vectors from $\mathbb{R}^n$ are thought of as ordered sets of three vectors from $\mathbb{R}^{n_i}$. In

[4]The desired coefficients describe the reflection of the waves, moving along the x-axis to the right with the maximal speed, from nonhomogeneities of the medium.

this case, the $n \times n$ matrices Λ and Q are naturally divided into the blocks: matrices $n_i \times n_j$. We denote them by Λ_{ij} and Q_{ij}. Denote by E_k the unit $n_k \times n_k$ matrix. Since the matrix Λ is diagonal, $\Lambda_{ij} = 0$ for $i \neq j$. The blocks Λ_{ii} are diagonal as well. Let

$$\Lambda_{00} = E_0, \qquad \Lambda_{11} = [a_1(x), a_2(x), \dots, a_{n_1}(x)] \stackrel{\text{def}}{=} A(x),$$

$$\Lambda_{22} = [-b_1(x), -b_2(x), \dots, -b_{n_2}(x)] \stackrel{\text{def}}{=} -B(x).$$

All the diagonal elements of the matrices A and B are positive; for all x, $a_j(x) < 1$ ($j = 1, 2, \dots, n_1$).

From the physical point of view, the conditions imposed on the matrix Λ mean that waves in the medium may propagate with various speeds. Besides, there is a separate group of waves which propagate to the right with the maximal speed equal to 1. We shall call these waves the rapid waves.

We shall consider below only matrix solutions of system (3.4.1): $U = (U_0, U_1, U_2)$ (the blocks U_k of the matrix U are aligned in the row to save room). Here U_k are $n_k \times n_0$ matrices. In these notations, system (3.4.1) takes the form

$$\begin{aligned} U_{0t} + U_{0x} &= Q_{00}U_0 + Q_{01}U_1 + Q_{02}U_2, \\ U_{1t} + AU_{1x} &= Q_{10}U_0 + Q_{11}U_1 + Q_{12}U_2, \\ U_{2t} - BU_{2x} &= Q_{20}U_0 + Q_{21}U_1 + Q_{22}U_2. \end{aligned} \tag{3.4.2}$$

First we set the direct problem. In the domain[5]

$$\mathcal{D}_l = \{x, t \, : \, t < l + \int_x^l (\max_j b_j(\xi))^{-1} \, d\xi\}$$

we seek for the matrix U satisfying the conditions

$$U|_{t<x} = 0, \tag{3.4.3}$$

$$\begin{aligned} (U_0 - R_{02}U_2)|_{x=0} &= E_0\delta(t) + \hat{H}_0(t), \\ (U_1 - R_{12}U_2)|_{x=0} &= \hat{H}_1(t). \end{aligned} \tag{3.4.4}$$

Here R_{ik} are $n_i \times n_k$ matrices; the matrices $\hat{H}_i(t)$ continuously depend on t for $t > 0$.

[5]Here and further in analogous formulas, maximum and minimum are taken over j.

3.4.3. Setting of the inverse problem. Formulation of the result

We shall consider the following inverse problem. Suppose that all the matrices $Q_{ik}(x)$ except for $Q_{20}(x)$ are known. It is required to find the matrix $Q_{20}(x)$ in the interval $(0,l)$ if the matrices $\hat{H}_0(t)$ and $\hat{H}_1(t)$ are given, and $\hat{H}_2(t) := U_2|_{x=0}$ is given in the interval $(0,T)$ where

$$T = \int_0^l \frac{d\xi}{\min b_j(\xi)}.$$

Thus, the matrices U_i satisfy the conditions (which are overdetermined from the point of view of the direct problem):

$$U_k = \delta_{k0}\delta(t) + H_k(t). \tag{3.4.5}$$

Here $H_k(t) = \hat{H}_k(t) + R_{k2}H_2(t)$, $k = 0, 1$.

The following statement holds:

1) The inverse problem has a unique solution.

2) For arbitrary given matrices, there exists l^* such that the inverse problem is solvable in the interval $(0,l^*)$.

3.4.4. Proof

The proof of this statement follows from the fact that the inverse problem is equivalent to a nonlinear system of integral Volterra equations of the second kind. We shall derive this system. To this end, we integrate each equation of system (3.4.1) along the corresponding characteristic of the system (x,t is an arbitrary point of the domain $\mathcal{D}_l$).

As a result, we obtain

$$U_0(x,t) = \int_0^x (Q_{00}(\xi)U_0(\xi, t - x + \xi) + Q_{01}U_1 + Q_{02}U_2)\, d\xi$$
$$+ H_0(t-x), \tag{3.4.6}$$

$$[U_1(x,t)]^{(j)} = \int_0^x \Big[Q_{10}(\xi)U_0\Big(\xi, t - \int_\xi^x \frac{d\xi'}{a_j(\xi')}\Big) + Q_{11}U_1 + Q_{12}U_2\Big]^{(j)} \frac{d\xi}{a_j(\xi)}$$
$$+ \Big[H_1\Big(t - \int_0^x \frac{d\xi}{a_j(\xi)}\Big)\Big]^{(j)}, \tag{3.4.7}$$

$$[U_2(x,t)]^{(j)} = -\int_0^x \Big[Q_{20}(\xi)U_0\Big(\xi, t + \int_\xi^x \frac{d\xi'}{b_j(\xi')}\Big) + Q_{21}U_1 + Q_{22}U_2\Big]^{(j)} \frac{d\xi}{b_j(\xi)}$$

$$+ \left[H_2\left(t + \int_0^x \frac{d\xi}{b_j(\xi)}\right)\right]^{(j)}. \tag{3.4.8}$$

In each of formulas (3.4.6)–(3.4.8) we indicate the arguments of the integrands only in the first term because in the next terms they are the same. The symbol $[\cdot]^{(j)}$ stands for the vector which is the j-th row of the corresponding matrix.

The function $U(x,t)$ has the following singularities for $t = x$:

$$U_0(x,t) = P_0(x)\delta(t-x) + O(1), \tag{3.4.9}$$

$$U_k(x,t) = P_k(x)\varepsilon(t-x) + O(t-x), \qquad k = 1, 2, \tag{3.4.10}$$

where the functions $P_i(x)$ satisfy the relations

$$\frac{d}{dx}P_0 = Q_{00}P_0, \qquad P_0|_{x=0} = E_0, \tag{3.4.11}$$

$$(E_1 - A)P_1 = Q_{10}P_0, \tag{3.4.12}$$

$$(E_2 + B)P_2 = Q_{20}P_0. \tag{3.4.13}$$

Equation (3.4.11) implies that the matrix P_0 is invertible. Relation (3.4.12) follows from equation (3.4.7). Taking into account that $U|_{t<x} = 0$ and U_0 has the δ-singularity, we can rewrite (3.4.7) as follows:

$$[U_1(x,t)]^{(j)} = \int_{x_j^*+0}^x \left[Q_{10}(\xi)U_0\left(\xi, t - \int_\xi^x \frac{d\xi'}{a_j(\xi')}\right) + Q_{11}U_1 + Q_{12}U_2\right]^{(j)} \frac{d\xi}{a_j(\xi)}$$
$$+ [\Phi_1(x,t)]^{(j)}. \tag{3.4.14}$$

Here

$$[\Phi_1(x,t)]^{(j)} = \begin{cases} H_1\left(t - \int_0^x \frac{d\xi}{a_j(\xi)}\right), & \text{if } t - \int_0^x \frac{d\xi}{a_j(\xi)} \geq 0, \\ (1 - a_j(x_j^*))^{-1}[Q_{10}(x_j^*)P_0(x_j^*)]^{(j)}, & \text{if } t - \int_0^x \frac{d\xi}{a_j(\xi)} < 0, \end{cases}$$

and x_j^* is the root of the equation

$$t - \int_{x^*}^x \frac{d\xi}{a_j(\xi)} = x^*.$$

The sign of the expression $t - \int_0^x \frac{d\xi}{a_j(\xi)}$ shows whether the corresponding characteristic intersects the "rapid" characteristic $t = x$; x_j^* is the coordinate of the point of intersection.

To obtain the last integral equation we solve (3.4.13) for Q_{20}:

$$Q_{20}(x) = (E_2 + B)P_2P_0^{-1} = (E_2 + B)U_2|_{t=x+0}P_0^{-1}, \tag{3.4.15}$$

and replace $U_2|_{t=x+0}$ by its expression from (3.4.8). The resulting system (3.4.6), (3.4.8), (3.4.14), (3.4.15) is a nonlinear Volterra system of second kind.

3.5. SECOND-ORDER HYPERBOLIC SYSTEM

3.5.1. Setting of a problem

We shall consider an inverse problem for the system of equations

$$C^{-2}U_{tt} - U_{xx} + QU = 0, \qquad x > 0. \tag{3.5.1}$$

For simplicity we restrict ourselves to the two-dimensional case: U, C, Q are 2×2 matrices. The matrix C is diagonal and constant. Its diagonal elements $c_1 = 1$, $c_2 = c$ $(0 < c < 1)$ are the velocities of the wave propagation. The greater velocity, without loss of generality, is assumed to be equal to unity. The matrices $Q = Q(x)$ and $U(x,t)$ are, generally speaking, complex-valued.

We suppose that $U|_{t<x} = 0$. The values of U and U_x are given for $x = 0$. It is convenient for us to introduce a function V as

$$U_x + C^{-1}U_t = V. \tag{3.5.2}$$

Then V satisfies the equation

$$V_x - C^{-1}V_t = QU. \tag{3.5.3}$$

Equations (3.5.2), (3.5.3) form a system of first-order hyperbolic equations. The values of the functions $U|_{x=0}$ and $V|_{x=0}$ are given for $t \in (0, 2T)$ as follows[6]:

$$U|_{x=0} = \delta(t)E + f(t), \tag{3.5.4}$$

[6]Evidently, if $U|_{x=0}$ is known, then it does not matter which of the functions $V|_{x=0}$ or $U_x|_{x=0}$ is given.

$$V|_{x=0} = g(t). \tag{3.5.5}$$

Here $f(t)$ and $g(t)$ are functions continuous for $t \in (0, 2T)$.

Suppose that we look for the function $Q(x)$ in the interval $(0, T)$. Let

$$l_1 = \frac{2c}{1+c}T, \qquad l_2 = cT \qquad (l_2 < l_1 < T),$$

$$\mathcal{D}_T = \{x, t \,:\, 0 < x < t < 2T - x\}.$$

The above inverse problem turns out to have the following properties. First, its solution is not unique. Second, if T is sufficiently small, then we can take arbitrary continuous functions $\alpha_{12}(x)$ and $\alpha_{21}(x)$ for $x \in (l_1, T)$, $\alpha_{22}(x)$ for $x \in (l_2, T)$ and set

$$\begin{cases} \begin{cases} q_{12}(x) = \alpha_{12}(x) \\ q_{21}(x) = \alpha_{21}(x) \end{cases} & \text{for} \quad x \in (l_1, T), \\ \quad q_{22}(x) = \alpha_{22}(x) & \text{for} \quad x \in (l_2, T). \end{cases}$$

Then the rest of the desired functions, namely, $q_{11}(x)$ for all $x \in (0, T)$, $q_{12}(x)$ and $q_{21}(x)$ for $x \in (0, l_1)$, and $q_{22}(x)$ for $x \in (0, l_2)$, are uniquely recovered.

Remark 3.5.1. It makes no sense to recover $Q(x)$ in the interval $(0, T_1)$ where $T_1 > T$. Really, since the maximal velocity is equal to unity, the functions $f(t)$ and $g(t)$, for $t < 2T$, provide no information on $Q(x)$ for $x > T$. At the same time, it is clear that when deriving equations for the inverse problem, we should take into account the behavior of the solution to the direct problem for all $x \in (0, T)$; otherwise we overlook the fact that $U = 0$ for $t < x$. Neglect of this fact may give rise to the waves which run from the right, get the boundary $x = 0$ until $t = 2T$ and influence the values of the inverse problem data $f(t)$ and $g(t)$.

3.5.2. Proof

The below proof is based on reducing the inverse problem to a system of nonlinear integral equations by a method analogous to those discussed in the previous sections. It should be noted that this system will not be of Volterra type.

Before deducing this system, we test the solution of the direct problem for singularities. It is easy to see that

$$U(x, t) = \begin{pmatrix} \delta(t - x) & 0 \\ 0 & \delta(t - x/c) \end{pmatrix} + u(x, t), \tag{3.5.6}$$

$$V(x,t) = v(x,t). \tag{3.5.7}$$

Here, we use small letters to denote matrices which are piecewise-continuous with respect to x and t. Note also that $u_{2k}|_{t=x+0} = 0$, where u_{ik} are the elements of the matrix $u(x,t)$. We proceed now to derive the system of integral equations. To this end, we integrate each of the differential equations of system the (3.5.2), (3.5.3) along the corresponding characteristic whose endpoints are located as follows: one is an arbitrary point of $\mathcal{D}_T$, another lies on that part of the boundary of $\mathcal{D}_T$ where $x = 0$ or $t = x$. Integrating (3.5.2), we obtain four equations

$$u_{ik}(x,t) = \int_{x^*}^{x} v_{ik}\Big(\xi, t + \frac{\xi - x}{c_i}\Big)\,\mathrm{d}\xi + \varphi_{ik}(x,t). \tag{3.5.8}$$

Here

$$x^* = \begin{cases} 0 & \text{for } \ i = 1, \\ \max\Big(0, \dfrac{x - ct}{1 - c}\Big) & \text{for } \ i = 2, \end{cases}$$

$$\begin{cases} \varphi_{ik}(x,t) = f_{ik}(t - x/c_i) & \text{for } \ t > x/c \quad (i = 1, 2), \\ \begin{cases} \varphi_{1k}(x,t) = f_{1k}(t - x/c_1) \\ \varphi_{2k}(x,t) = 0 \end{cases} & \text{for } \ t < x/c. \end{cases}$$

The equations of the first group are illustrated schematically in Fig. 3.2. The segments of the lines passing through the point (x,t) represent the characteristics along which we integrate.

To obtain the second group of equations, we turn to equation (3.5.3). Note that the expression in the right-hand side of (3.5.3) may have δ-singularities concentrated along the characteristics $t = x$ or $t = x/c$ (see formula (3.5.6)). We shall try to choose the characteristic, along which we integrate, so that it does not pass through the δ-singularity of the integrand and does not end at it.

Then, for $v_{11}(x,t)$, $(x,t) \in \mathcal{D}_T$, we get

$$v_{11}(x,t) = \int_0^x (q_{11}u_{11} + q_{12}u_{21})(\xi, t + x - \xi)\,\mathrm{d}\xi + g_{11}(0, t + x). \tag{3.5.9a}$$

Suppose that $t + x/c < 2T$; then

$$v_{21}(x,t) = \int_0^x (q_{21}u_{11} + q_{22}u_{21})(\xi, t + (x - \xi)/c)\,\mathrm{d}\xi + g_{21}(t + x/c). \tag{3.5.9b$_1$}$$

For $t + x/c > 2T$, the characteristic along which the integration is carried out in (3.5.9a) intersects the boundary of the domain $\mathcal{D}_T$ at the point

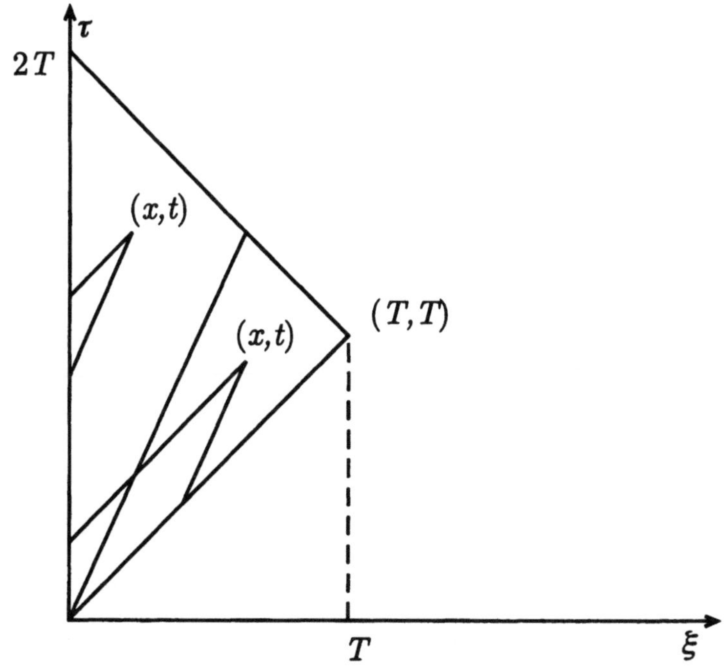

Fig. 3.2

$\left(\dfrac{x+ct-2cT}{1-c}, \dfrac{2T-x-ct}{1-c}\right)$. Since we have no information on the value of v_{21} at this point, we shall integrate the equation corresponding to the second row and the first column in (3.5.3) along the characteristic which passes through (x,t) exactly as in (3.5.9b$_1$) but in the opposite direction. Then, taking into account that under the integral sign there is a term which contains the δ-singularity, we obtain

$$v_{21}(x,t) = -\int_x^{(x+ct)/(1+c)} (q_{21}u_{11} + q_{22}u_{21})(\xi, t + (x-\xi)/c)\,\mathrm{d}\xi \\ - \frac{c}{1+c} q_{21}\left(\frac{x+ct}{1+c}\right). \tag{3.5.9b$_2$}$$

The pair of equalities (3.5.9b$_1$) and (3.5.9b$_2$) represent the second equation which takes various forms depending on the position of the point (x,t) (see Fig. 3.3).

For the equations of the second column of (3.5.3), the δ-singularities of their right-hand sides are located on the characteristic $t = x/c$. Therefore, we shall look for characteristics not crossing $t = x/c$ to integrate along them.

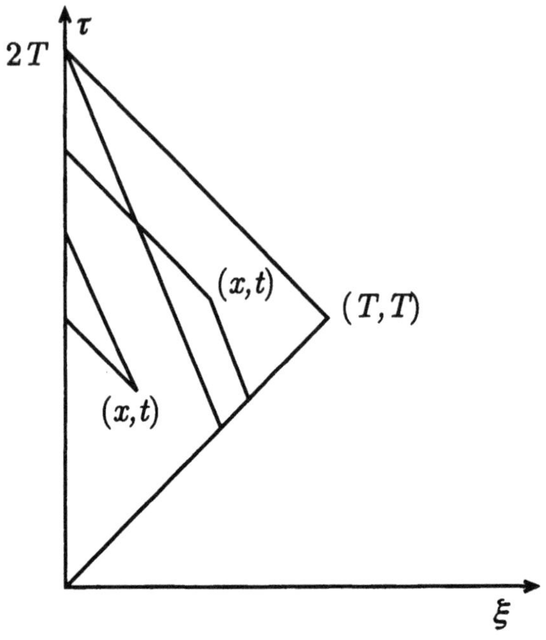

Fig. 3.3

The equation of the first row and the second column of (3.5.3), for $t > x/c$, is integrated along the characteristic which connects the point (x,t) with a point in the interval $(0, 2T)$ of the t-axis, resulting in

$$v_{12}(x,t) = \int_0^x (q_{11}u_{12} + q_{12}u_{22})(\xi, t + x - \xi)\,\mathrm{d}\xi + g_{12}(t+x). \qquad (3.5.9\mathrm{c}_1)$$

For $t < x/c$, we integrate along the same characteristic but in the opposite direction to derive

$$v_{12}(x,t) = -\int_x^{(x+t)/2} (q_{11}u_{12} + q_{12}u_{22})(\xi, t + x - \xi)\,\mathrm{d}\xi. \qquad (3.5.9\mathrm{c}_2)$$

Finally, the equation of the second row and the second column, for $x/c < t < 2T - x/c$, is integrated along the characteristic connecting the point (x,t) with a point in the interval $(0, 2T)$ of the t-axis. This yields

$$v_{22}(x,t) = \int_0^x (q_{21}u_{12} + q_{22}u_{22})(\xi, t + (x-\xi)/c)\,\mathrm{d}\xi + g_{22}(t + x/c). \quad (3.5.9\mathrm{d}_1)$$

For $t < x/c$, we have

$$v_{22}(x,t) = -\int_x^{(x+ct)/(1+c)} (q_{21}u_{12} + q_{22}u_{22})(\xi, t + (x-\xi)/c)\,\mathrm{d}\xi. \quad (3.5.9\mathrm{d}_2)$$

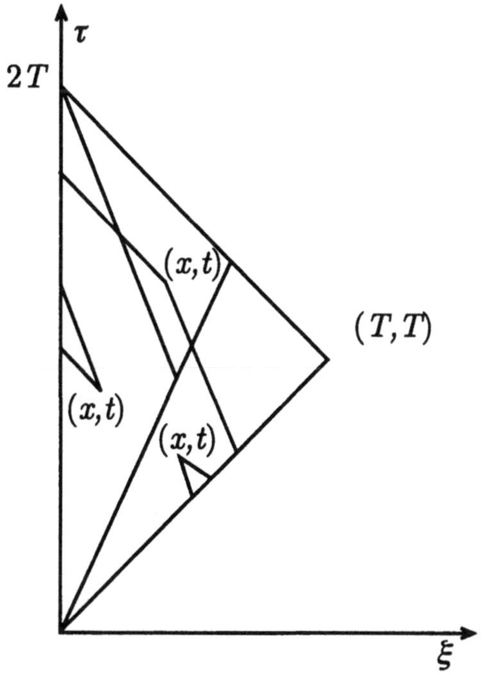

Fig. 3.4

Finally, for $t > \max(x/c, 2T - x/c)$, we cannot find a characteristic not intersecting $t = x/c$. In this case, the equation resulted from the integration differs from (3.5.9d$_2$) in having an additional term (because of the δ-term):

$$v_{22}(x,t) = -\int_x^{(x+ct)/(1+c)} (q_{21}u_{12} + q_{22}u_{22})(\xi, t + (x-\xi)/c)\,d\xi$$

$$- \frac{c}{2} q_{22}((x+ct)/2). \tag{3.5.9d$_3$}$$

Equations (3.5.9a–d) form the second group of integral equations. The third group will be obtained if we integrate each equation of system (3.5.3) along a characteristic connecting a point in the interval $(0, 2T)$ of the t-axis with a point lying under the characteristic $t = x$. Taking into account that $v = 0$ for $t < x$, we obtain, for an arbitrary $t \in (0, 2T)$,

$$g_{11}(t) + \int_0^{t/2} (q_{11}u_{11} + q_{12}u_{21})(\xi, t-\xi)\,d\xi + \frac{1}{2} q_{11}(t/2) = 0,$$

$$g_{21}(t) + \int_0^{ct/(1+c)} (q_{21}u_{11} + q_{22}u_{21})(\xi, t - \xi/c)\,d\xi + \frac{c}{1+c} q_{21}(ct/(1+c)) = 0,$$

$$g_{12}(t) + \int_0^{t} (q_{11}u_{12} + q_{12}u_{22})(\xi, t - \xi)\,d\xi + \frac{c}{1+c} q_{12}(ct/(1+c)) = 0,$$

$$g_{22}(t) + \int_0^{ct/(1+c)} (q_{21}u_{12} + q_{22}u_{22})(\xi, t - \xi/c)\,d\xi + \frac{c}{2} q_{22}(ct/2) = 0.$$

The integrated terms in the last four formulas appear because the characteristic along which we integrate intersects the carrier of the δ-term. Now, in each of the four equalities we denote by x the argument of the function q_{ik} in the last term. Then, after rearranging, we obtain

$$q_{11}(x) = -2\int_0^{x} (q_{11}u_{11} + q_{12}u_{21})(\xi, 2x - \xi)\,d\xi - 2g_{11}(2x), \quad x \in (0, T), \tag{3.5.10a}$$

$$q_{21}(x) = -\frac{1+c}{c}\int_0^{x} \Big(q_{21}u_{11} + q_{22}u_{21}\Big)\Big(\xi, \frac{1+c}{c}x - \frac{\xi}{c}\Big)\,d\xi - \frac{1+c}{c} g_{21}\Big(\frac{1+c}{c}x\Big), \quad x \in \Big(0, \frac{2cT}{1+c}\Big), \tag{3.5.10b}$$

$$q_{12}(x) = -\frac{1+c}{c}\int_0^{(1+c)x/(2c)} \Big(q_{11}u_{12} + q_{12}u_{22}\Big)\Big(\xi, \frac{1+c}{c}x - \xi\Big)\,d\xi - \frac{1+c}{c} g_{12}\Big(\frac{1+c}{c}x\Big), \quad x \in \Big(0, \frac{2cT}{1+c}\Big), \tag{3.5.10c}$$

$$q_{22}(x) = -\frac{2}{c}\int_0^{cx/(1+c)} \Big(q_{21}u_{12} + q_{22}u_{22}\Big)\Big(\xi, \frac{2x - \xi}{c}\Big)\,d\xi - \frac{2}{c} g_{22}\Big(\frac{2x}{c}\Big), \quad x \in (0, cT) \tag{3.5.10d}$$

(see Fig. 3.5, 3.6).

Throughout equations (3.5.8)–(3.5.10) we replace the functions $q_{12}(x)$, $q_{21}(x)$ for $x \in (l_1, T)$ and $q_{22}(x)$ for $x \in (l_2, T)$ by $\alpha_{12}(x)$, $\alpha_{21}(x)$, and $\alpha_{22}(x)$, respectively.

Introduce a Banach space C_T whose elements are the sets X of 12 functions:

$$X_{ik}^{(1)}(x,t) = u_{ik}(x,t), \qquad X_{ik}^{(2)}(x,t) = v_{ik}(x,t), \qquad (x,t) \in \mathcal{D}_T;$$

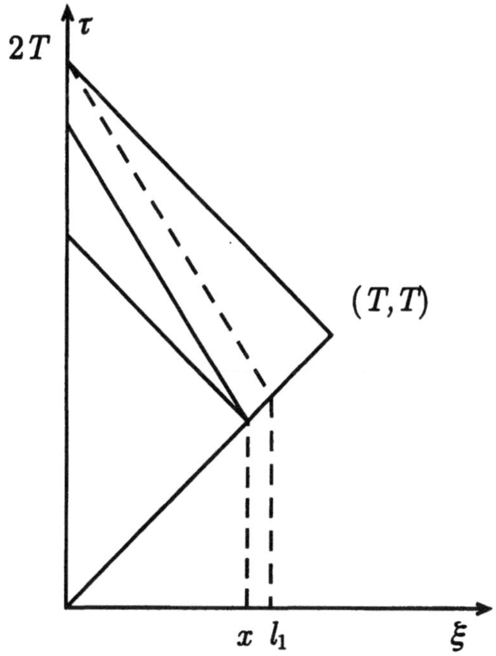

Fig. 3.5

$$X_{11}^{(3)}(x) = q_{11}(x), \quad x \in (0,T); \quad X_{ik}^{(3)}(x) = q_{ik}(x) \quad (i \neq k), \quad x \in (0,l_1);$$

$$X_{22}^{(3)}(x) = q_{22}(x), \quad x \in (0,l_2).$$

The functions $X_{ik}^{(3)}(x)$ are continuous, the functions $X_{ik}^{(j)}(x,t)$ $(j = 1,2)$ are piecewise-continuous in $\mathcal{D}_T$ (discontinuities may occur on the characteristic $t = x/c$). The norm in C_T is defined as

$$\|X\| = \max\Big(\sup_{\substack{ik \\ (x,t)\in\mathcal{D}_T}} (|X_{ik}^{(1)}(x,t)|, |X_{ik}^{(2)}(x,t)|), \sup_{x\in(0,T)} |X_{11}^{(3)}(x)|,$$

$$\sup_{\substack{i\neq k \\ x\in(0,l_1)}} |X_{ik}^{(3)}(x)|, \sup_{x\in(0,l_2)} |X_{22}^{(3)}(x)|\Big).$$

Then, given continuous functions $\alpha_{12}(x)$, $\alpha_{21}(x)$, and $\alpha_{22}(x)$, the system (3.5.8)–(3.5.10) can be written in the form

$$X = KX, \tag{3.5.11}$$

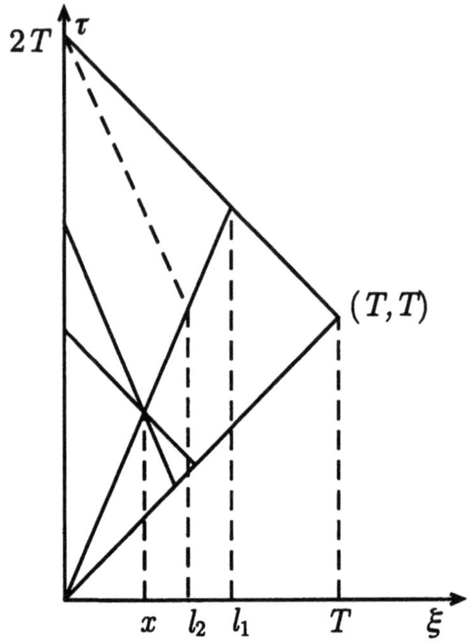

Fig. 3.6

where K is an operator $C_T \to C_T$. It is easy to show that if $\|X\| \leq M$, $|f_{ik}(t)| \leq m$, $|g_{ik}| \leq m$, $|\alpha_{ik}| \leq m$, then $\|KX\| \leq 2/c(2(M+m)Ml+m)$. If $\|X\| \leq M$ and $\|\tilde{X}\| \leq M$, then $\|KX - K\tilde{X}\| \leq 4l/c(2M+m)\|X-\tilde{X}\|$. We choose the constants M and l so that

1) $2/c(2(M+m)Ml+m) \leq M$; then the operator K maps the ball of radius M centered at zero of the space C_T into itself;

2) $4l/c(2M+m) < 1$; then the operator K is a contraction operator.

Evidently, such M and l exist. By the Banach theorem, equation (3.5.11) is uniquely solvable. Equation (3.5.11) is equivalent to the inverse problem (3.5.2)–(3.5.5). Thus, the solution of the inverse problem (3.5.2)–(3.5.5) is not unique. The class of possible solutions is completely parametrized by the triple of functions α_{12}, α_{21}, α_{22}.

3.5.3. The inverse problem with a fixed interval of nonhomogeneities

As before, we consider the inverse problem (3.5.2), (3.5.3) with the data

(3.5.4), (3.5.5), but now we suppose that

$$Q(x)|_{x>l} = 0. \tag{3.5.12}$$

Conditions (3.5.4), (3.5.5) are set in the interval $(0, 2T)$, where $T = l/c$. The following statement holds.

Proposition 3.5.1. The matrix $Q(x)$ is recovered uniquely from the functions $f(x)$, $g(x)$ on the interval $(0, l)$ if l is sufficiently small.

Proposition 3.5.1 follows, in fact, from the reasoning of Subsection 3.5.2. Really, as shown in the previous subsection, the matrix $Q(x)$ is recovered uniquely in the interval $(0, T)$ from the functions $f(x)$, $g(x)$ for a sufficiently small T, if there is an additional information on behavior of the desired matrix in the interval (l, T) (that is, the functions $\alpha_{12}(x)$, $\alpha_{21}(x)$, $\alpha_{22}(x)$ are given). In this case, condition (3.5.12) means that $Q(x)$ and therefore $\alpha_{12}(x)$, $\alpha_{21}(x)$, $\alpha_{22}(x)$ vanish in the interval (l, T). At the same time, condition (3.5.12) is essentially stronger than the requirement that $\alpha_{12}(x)$, $\alpha_{21}(x)$, $\alpha_{22}(x)$ be equal to zero. Therefore, the problem in question may have no solution. It turns out that the theorem of unique solvability holds for a sufficiently small l if the functions $f_{k1}(t)$, $g_{k1}(t)$ are given in the interval $(0, l+l/c)$, and the functions $f_{k2}(t)$, $g_{k2}(t)$, as before, are given in the interval $(0, 2l/c)$. To prove this statement, we reduce the problem to a system of integral equations relative to the twelve functions $u_{ik}(x,t)$, $v_{ik}(x,t)$, $q_{ik}(x)$. Here $u_{i1}(x,t)$, $v_{i1}(x,t)$ are defined in the domain $\mathcal{D}_{l1} = \{x, t : 0 < x < t < (1/c)(l(1+c) - x),\ x < l\}$; $u_{i2}(x,t)$, $v_{2k}(x,t)$, in the domain $\mathcal{D}_{l2} = \{x, t : 0 < x < t < (1/c)(2l - x),\ x < l\}$; and $q_{ik}(x)$, in the interval $(0, l)$. The domains $\mathcal{D}_{l1}$, $\mathcal{D}_{l2}$ are shown in Fig. 3.7

We now construct these equations. The first four equations coincide formally with equations (3.5.8) with the only difference that, for $k = 1$, $(x,t) \in \mathcal{D}_{l1}$ and, for $k = 2$, $(x,t) \in \mathcal{D}_{l2}$. The next pair of equations is as follows:

$$v_{i1}(x,t) = \int_0^x (q_{i1}u_{i1} + q_{i2}u_{21})(\xi, t + (x-\xi)/c_i)\,\mathrm{d}\xi + g_{i1}(t + x/c_i), \quad (x,t) \in \mathcal{D}_{l1}. \tag{3.5.13}$$

The next pair of equations for $(x,t) \in \mathcal{D}_{l2}$ is represented by four equalities, namely: for $t > x/c$,

$$v_{i2}(x,t) = \int_0^x (q_{i1}u_{12} + q_{i2}u_{22})(\xi, t + (x-\xi)/c_i)\,\mathrm{d}\xi + g_{i2}(t + x/c_i) \tag{3.5.14a}$$

Belichev, M. I., Blagoveshchenskii A. S., and Ivanov, S. (1997). Two-velocity dinamical system. *Boundary Control of Waves Inverse Problems. Wave Motion*, 28 (1), 83–107.

Blagoveshchenskii, A. S. (1966). On the inverse problem of seismic wave propagation theory. *Problems of Math. Physics*, 1, Leningrad. Univ. Press, Leningrad, 68–81 (in Russian).

Blagoveshchenskii, A. S. (1969). One-dimensional inverse boundary-value problem for the hyperbolic equation of the second order. *Zapiski Nauchn. Seminarov LOMI*, 78, 85–90 (in Russian).

Blagoveshchenskii, A. S. (1970). On various setting of the one-dimensional inverse problem for telegraph equation. *Problems of Mathemat. Physics.* 4, Leningrad. Univ. Press, Leningrad, 40–41 (in Russian).

Blagoveshchenskii, A. S. (1971). On a local method of solution of the non-stationary inverse problem for the inhomogeneous string. *Trudy Mat. Inst. im. Steklova*, CXV, Nauka, Leningrad, 28–38 (in Russian).

Blagoveshchenskii, A. S. (1979a). An inverse problem of wave propagation in a stochastic medium. *Zapiski Nauchn. Seminarov LOMI*, 89, 63–70 (in Russian).

Blagoveshchenskii, A. S. (1979b). Inverse problems for systems of hyperbolic type. *Problems of Math. Physics*, Leningrad. Univ. Press. Leningrad, 3–14 (in Russian).

Blagoveshchenskii, A. S. (1983). On the existence theorem for inverse problem of the theory of wave propagation in a layered medium. *Theory and Methods of Solution of Ill-Posed Problems and their Application*, Novosibirsk, 44–45 (in Russian).

Blagoveshchenskii, A. S. (1986). Inverse problems of acoustic in a moving medium. *Problems of Mathem. Physics*, 11, Leningrad. Univ. Press., Leningrad, 46–58 (in Russian).

Blagoveshchenskii, A. S. (1992). An inverse axial-symmetric Lemb problem. *Zapiski Nauchn. Seminarov LOMI*, 203, 51–67 (in Russian).

Blagoveshchenskii, A. S. (1995). On the approach to inverse problem of the wave propagation in dissipative layered media. *PDMI Preprint No.8.*

Blagoveshchenskii, A. S. and Lavrent'ev, K. K. (1975). Inverse problem of determination of the boundary condition in the theory of propagation of nonstationary waves. I. *Zapiski Nauchn. Seminarov LOMI*, 51, 78–84 (in Russian).

Blagoveshchenskii, A. S. and Voevodskii, K. E. (1981). Inverse problem of the wave scattering form layered-inhomogeneous half space. *Differential Equations*, XVII (8), 1434–1445 (in Russian).

Blokhintsev, D. I. (1981). *Acoustic of inhomogeneous Moving Medium.* Nauka, Moscow (in Russian).

Gardner, C. S., Green, J. M., Kruskal, M. D., and Miura, R. M. (1967). Method for solving the Korteweg-de Vries equation. *Phys. Rev. Lett*, 19 (19), 1095–1097.

Gel'fand, I. M. and Levitan, B. M. (1951). On determination of the differential equation by its spectral function. *Izv. Acad. Nauk. SSSR, ser. Math.*, 15 306–360 (in Russian).

Gel'fand, I. M. and Shilov, G. E. (1959). *Generalized Functions and Actions with them.* Nauka, Moscow (in Russian).

Goryunov, A. A. and Saskovets, A. V. (1989). *Inverse Scattering Problems in Acoustic.* Mosk. Gos. Univer., Moscow (in Russian).

Kabanikhin, S. I. and Lorenzi, A. (1999). *Identification Problems of Wave Phenomena.* VSP, Utrecht.

Kay, I. and Moses, H. E. (1956). The determination of the scattering potential from the spectral measure function. *Nuovo Cimento*, 3 (2), 276–304.

Kolmogorov, A. N. and Fomin, S. V. (1981). *Elements of the Function Theory and Functional Analysis.* Nauka, Moscow (in Russian).

Krein, M. G. (1955). On the method of the effective solution of the inverse bounary-value problem. *Dokl. Akad. Nauk SSSR*, 94 (6), 987–990 (in Russian).

Lavrent'ev, K. K. (1975). Inverse problem of determination of the boundary condition in the theory of propagation of nonstationary waves. II. *Zapiski Nauchn. Seminarov. LOMI*, 51, 129–133 (in Russian).

Lavrent'ev, M. M., and Romanov, V. G. (1966). Three linearized inverse problems for hyperbolic equations. *Dokl. Acad. Nauk. SSSR*, 171 (6), 1279–1281 (in Russian).

Lavrent'ev, M. M., Romanov, V. G. and Vasil'ev, V. G. (1969). *Multidimensional Inverse Problems for Differential Equations.* Nauka, Novosibisk (in Russian).

Lavrenti'ev, M. M., Romanov, V. G., and Shishatskii, S. P. (1986). *Ill-Posed Problems of Mathematical Physics and Analysis.* American Mathematical Society, Providence, Rhode Island.

Marchenko, V. A. (1955). Recovery of potential energy by the phases of scattering waves. *Dokl. Akad. Nauk SSSR*, 104 (5), 350–357 (in Russian).

Megrabov, A. G. (1972). Direct and inverse problems of scattering plane waves in the inhomogeneous transient layers. *Mathem. Problems of Geophysics*, 3, Novosibirsk, 113–123 (in Russian).

Megrabov, A. G. (1973). Inverse problems for the elliptic equation in a strip connected with the problem of the plane waves scattering in the inhomogeneous layers. *Mathem. Problems. of Geophysics*, 4, Novosibirsk, 103–115 (in Russian).

Nizhnik, L. P. (1991). Inverse Problems of Scattering for Hyperbolic Equations. Naukova Dumka, Kiev (in Russian).

Nizhnik, L. P. and Tarasov, V. G. (1977). An inverse nonstationary problem of scattering for hyperbolic equations system. *Dokl. Akad. Nauk. SSSR*, 233 (3), 300–303 (in Russian).

Pariiskii, B. S. (1969). The inverse problem for the wave equation with the interaction in the depth. *Calculating Seismology*, 2, Nauka, Moscow, 37–51 (in Russian).

Romanov, V. G. (1976). On an inverse problem for weakly connected hyperbolic systems of the first order. *Ill-Posed Mathematical Problems and Problems of Geophysics.* Novosibirsk, 135–148 (in Russian).

Romanov, V. G. (1983). The inverse Lemb problem in linear approximation. *Numerical Methods in Seismic Researches*, Nauka, Novosibirsk, 170–192 (in Russian).

Romanov, V. G. (1987). *Inverse Problems of Mathematical Physics.* VNU Science Press, Utrecht, The Netherlands.

Romanov, V. G. and Belinskii, S. P. (1976). *To the Problem of Determination of Coefficients of t-Hyperbolic System.* Computer Center, Siberian Branch of USSR Academy of Sciences, Preprint No. 23, Novosibirsk, 16–24 (in Russian).

Romanov, V. G. and Kabanikhin, S. I. (1991). *Inverse Problems of Geoelectrics.* Nauka, Moscow (in Russian).

Romanov, V. G. and Kabanikhin, S. I. (1994). *Inverse Problems of Maxwell's Equations.* VSP, Utrecht.

Romanov, V. G. and Slinucheva, L. I. (1972). An inverse problem for linear hyperbolic systems of the first order. *Mathematical Problems of Geophysics,* 3, Novosiborsk, 187–215 (in Russian).

Tikhonov, A. N. and Arsenin, V. Ya (1977). *Solution of Ill-Posed Problems.* Wiley, New York.

Yakhno, V. G. (1985). *One-Dimensional Inverse Dynamical Problems for Anisotropy Elactic Media.* Computer Center, Siberain Branch of USSR Academy of Science, Novosibirsk (in Russian).

Yakhno, V. G. (1990). *Inverse Problems for Differential Elastic Equations,* Nauka, Novosibirsk (in Russian).

Bei Fragen zur Produktsicherheit wenden Sie sich bitte an:
If you have any questions regarding product safety,
please contact:

Walter de Gruyter GmbH
Genthiner Straße 13
10785 Berlin
productsafety@degruyterbrill.com